SYNTHESE LIBRARY

STUDIES IN EPISTEMOLOGY,
LOGIC, METHODOLOGY, AND PHILOSOPHY OF SCIENCE

VOLUME 180

SOCIOBIOLOGY AND EPISTEMOLOGY

Edited by

JAMES H. FETZER

University of Virginia

D. REIDEL PUBLISHING COMPANY

A MEMBER OF THE KLUWER ACADEMIC PUBLISHERS GROUP

DORDRECHT / BOSTON / LANCASTER

Library of Congress Cataloging in Publication Data

Main entry under title:

Sociobiology and epistemology.

(Synthese library)
Includes bibliographies and indexes.
1. Sociobiology--Addresses, essays, lectures.
2. Knowledge, Theory of--Addresses, essays, lectures.
3. Science--Philosophy--Addresses, essays, lectures.
I. Fetzer, James H., 1940- .
GN365.9.S595 1985 304.5 85-14406
ISBN 90-277-2005-3
ISBN 90-277-2006-1 (Pallas paperback : pbk.)

Published by D. Reidel Publishing Company,
P.O. Box 17, 3300 AA Dordrecht, Holland.

Sold and distributed in the U.S.A. and Canada
by Kluwer Academic Publishers,
190 Old Derby Street, Hingham, MA 02043, U.S.A.

In all other countries, sold and distributed
by Kluwer Academic Publishers Group,
P.O. Box 322, 3300 AH Dordrecht, Holland.

Printed in The Netherlands

to E. O. Wilson

TABLE OF CONTENTS

EPILOGUE

FOREWORD

The papers presented in this special collection focus upon conceptual, theoretical and epistemological aspects of sociobiology, an emerging discipline that deals with the extent to which genetic factors influence or control patterns of behavior as well as the extent to which patterns of behavior, in turn, influence or control genetic evolution. The Prologue advances a comprehensive account of the field of gene-culture co-evolution, where Lumsden and Gushurst differentiate between "classical" sociobiology (represented especially by Wilson's early work) and current research on human sociobiology (represented by Lumsden and Wilson's later work), which emphasizes interplay between genes, minds, and culture. The specter of genetic determinism, no doubt, has created considerable controversy, some of which may be laid to rest by Hanna's analysis of the (ambiguous) notion of a "genetic program", which indicates the necessity for distinguishing between descriptive and prescriptive dimensions of this complex concept. Brandon offers a framework for assessing the respective contributions of nature and of nurture by advancing a means for measuring genetic and cultural influences upon "inheritance", which supports the conclusion that evolving patterns of behavior do not always maximize inclusive fitness, contrary to what sociobiologists have claimed. The influence of culture upon genetic evolution, of course, can be adequately appraised only when a suitable account of culture itself has been found, a desideratum Smillie attempts to satisfy by utilizing the notion of "cinfo" as culturally transmitted ecological information, a resource other species tend not to exploit. Rather than examine sociobiology from the perspective of philosophy, Masters suggests that modern "social biology" has significant implications for a broad range of traditional philosophical issues, including questions of justice and morality, a position that is reinforced by Thomas' account of parental love within the context of evolutionary theory, which not only provides a foundation for "altruism" but also hints that biology and morality might be allies rather than adversaries.

Part II begins with three papers discussing areas of sobiobiological theory-construction that appear to warrant further investigation. Richardson, for example, contends that controversies over the "units" and the "levels" of selection are unlikely to be resolved without access to detailed models of genetic/cultural evolution: the hypothesis that genes are the units of selection, in particular, is a difficult thesis to test. Rosenberg similarly suggests that, just as relevant features of biological evolution cannot be ascertained without extensive research, the relevant features of sociobiological evolution cannot be ascertained without extensive research – for it would be a mistake to uncritically presuppose that the patterns of behavior to which sociobiology applies correspond to those specified by the typology of "folk psychology". And Sober maintains that, while the assumption of methodological behaviorism underlies specific sociobiological arguments, certain models of cultural evolution, and various aspects of the theory of games, the deficiencies that may exist in these approaches, nevertheless, seem to be located elsewhere. The following two papers offer differing analyses of the logical structure and of the epistemic standing of sociobiological theory. Thompson, for example, embraces the "semantic conception" of scientific theories, which, he contends, provides a more accurate representation of the relations between a theory and its domain, affording a more adequate understanding of varied facets of sociobiological theorizing; while Fetzer, by contrast, endorses a conception of theories first advanced by Campbell, which, in effect, suggests that the central problem is not the structure of theories but the character of laws, especially laws relating genes, bodies, and minds. The Epilogue concludes with a provocative paper by Ruse, who urges that the conception of "epigenetic rules" has the capacity to illuminate important problems in the theory of knowledge (by extrapolating from sociobiology to epistemology), yet who disagrees with Masters by contending that Hume should properly qualify as Wilson's precursor, after all.

30 January 1985 J. H. F.

PROLOGUE

CHARLES J. LUMSDEN AND ANN C. GUSHURST

GENE-CULTURE COEVOLUTION: HUMANKIND IN THE MAKING

ABSTRACT. Recent evidence indicates that biological evolution and cultural history are linked together. Study of the linkage has been active in the past several years and constitutes a new phase of investigation in sociobiology. The work is of particular interest to philosophical and interdisciplinary inquiry because it avoids the earlier biological determinism and animal comparisons, emphasizing instead the multiple roles played by cultural transmission, genetic inheritance, and individual mental development within specifically human social systems. Culture is proposed to be shaped in part by biological mechanisms that direct the assembly of the human mind during socialization. These mechanisms create the rules by which human nature governs itself, supporting individual choice and the social forms emergent in human group behavior. The frequencies of the gene variants prescribing these mechanisms are in turn influenced by evolutionary forces exerted in the context of particular cultures. The forces include but are not limited to natural selection. Mathematical theories have been developed that begin to trace the course of the gene-culture coevolutionary relationship and use it to further unify biology with the social sciences and humanities.

INTRODUCTION

Human beings were not created by biological Darwinian evolution. For the past several million years our ancestors have been shaped by biological evolution and cultural evolution proceeding together in a manner still little understood. In each of 200,000 or so generations two tracks of heritable information, one genetic and the other cultural, have met in the events of individual socialization. During this time biology does not appear to have overwhelmed culture and culture has not granted biology total independence. The relationship is instead one of reciprocating interaction, in which culture is generated and shaped by biological imperatives while the course of biological evolution shifts in response to cultural opportunities. The analysis of the interaction is bringing fundamental issues in epistemology and the philosophy of mind into mainstream evolutionary biology and sociobiology.

Classical sociobiology (Wilson, 1975; Dawkins, 1976; Alexander, 1979) has previously attempted to link genes to social phenomena in a direct manner. Using the basic principles of population genetics and ecology, researchers have sought to explain and predict the environmental conditions

James H. Fetzer (ed.), Sociobiology and Epistemology, 3–28.

under which dominance systems, altruism, pair bonding, parental care, homosexuality, and other forms of social behavior are most likely to arise when individuals or groups optimize various types of genetic fitness. This method has been notably successful in the study of animals, particularly with regard to their relatively invariant, "instinctive" behaviors. It has been successful only to a limited degree in the analysis of human behavior. It has, for instance, not accounted for the activities of human (or animal) thought, consciousness, advanced emotion, and decision-making. Nor has it begun to deal adequately with the existence of culture or delineated even crudely the nature of the relationship between genetic and cultural evolution. Classical sociobiology operates primarily with fitness optimization rules inferred from theories of biological evolution, and a data base stocked with information about animal societies. What has been missing until recently is an attempt to build equally useful treatments of populations that evolve by the inheritance of both genetic and cultural information. There has also been little work on incorporating the extensive information about human neurobiology, psychology, and cultural anthropology into the sociobiological data base. Thus most scenarios about human evolution apply natural selection ideas based on straight genetic inheritance of behavioral traits, assuming that results derived from this process cover gene-culture interaction. When genes and culture interact, however, evolutionary processes much more complex than optimization of traits may occur (Lumsden and Wilson, 1981a; pp. 298ff).

The properties of the gene-culture linkage therefore pose problems not previously confronted by students of organic evolution and to which the new gene-culture theories represents tentative first answer. Critics (reviewed in Lumsden and Wilson, 1983) correctly point out that a sociobiology relevant to the social sciences and humanities must have expanded covering power, with the capacity to generate causal explanatory descriptions of mind, culture, and large-scale social patterns in addition to the population parameters previously of interest to biologists. A formal scientific theory is required that analyzes the social sciences from the viewpoint of modern evolutionary biology. This theory would explain the mechanisms responsible for the evolution of human nature and the linkage between genes and culture. It would also reconstruct patterns of cultural diversity in a manner that clearly depicts their relations to underlying biological properties. In response there now exists a growing body of data and theory pertaining to the relationship between biology and culture. Scholars have focussed on a number of basic issues, including the general qualitative properties of cultural change as constrasted with genetic change (Blum, 1963; Garn, 1964; Tinbergen, 1972); the similarity in the outcomes of genetic and cultural evolution (Emlen, 1976; Durham, 1979); the sometimes competitive nature of the two modes of change (Richerson and Boyd, 1978); the potential of the coevolutionary process to shape learning capacity (Feldman and Cavalli-Sforza, 1976; Pulliam and Dunford, 1980); the strength of the feedforwards from genes to culture and the feedbacks from culture to genes under various circumstances (Durham, 1979, 1982; Lumsden and Wilson, 1981a, 1983; Lumsden, 1983a, b); the complex and multiple nature of cultural transmission systems (Cavalli-Sforza and Feldman, 1981); the properties of mental activity and psychological development sustained by gene-culture interaction (Lumsden and Wilson, 1981a, 1983; Lumsden, 1983a, b, c, 1984a, b; Lopreato, 1984); and the zoological diversity of dual inheritance systems (Bonner, 1980; Lumsden and Wilson, 1981a; Lumsden, 1984c). These efforts have been of sufficiently general interest to produce a steadily expanding secondary literature devoted to the dissemination, popularization, ideological critique, and philosophical analysis of gene-culture theories (e.g., Lewontin, 1981; Lumsden and Wilson, 1981b; Lumsden and Wilson, and twenty-three respondents, 1982; Medawar, 1981; Trigg, 1982; Dobzhansky and Boesiger, 1983; Gould, 1983; Ortner, 1983; Swanson, 1983; Maxwell, 1984; Wilson and Lumsden, 1984).

The literature on the subject of gene-culture coevolution is already too extensive to be encompassed within a brief article. We hope to convey the general character of the coevolutionary theories now under development, and where it exists to identify emerging agreement about basic issues. Our characterization will of necessity be selective, and our biases will emphasize the experience gained in our own research. We request the forebearance of those colleagues whose contributions are inadvertently overlooked in our citations.

THE SUBSTANCE OF CULTURE

In exercising their passions, human beings make choices. The alternative behaviors and patterns of thought are not determined genetically. The mind develops them from information it acquires during socialization. This information is the substance of culture. Although the details vary (Kroeber and Kluckhohn, 1963), there is some agreement among anthropologists that culture is an ordered system of socially transmissible information encoding behavioral and cognitive characteristics of a social group, including such aspects as skills, knowledge, attitudes, beliefs, myths, and rituals (Kroeber and Kluckhohn, 1963; Schwartz and Ewald, 1968; Schneider and Bonjean, 1973; Keesing, 1976). This description of culture is often extended to include the actual lifeways and material artifacts that are the overt expressions of the transmissible information system. In this report we will mainly be concerned with the informational aspects *per se*.

An extensive literature affirms that human cultures, as information systems, possess an inner patterning (for overviews see Murdock, 1967; Naroll and Cohen, 1970; Taylor, 1973). The discourse and body movements involved in tasks, rituals, demonstrations and other means of transmission can be hierarchically clustered as sets of phonemes, words, sentences, stories and analogous components of motor patterns (Hutchinson, 1970; Laban, 1975). Considerable attention has focussed on these elements of these various sets as natural units or building blocks of culture. In culture theory previous authors have referred to units of culturally transmitted information in different ways, such as mnemotype, idea, idene, sociogene, instruction, culture type, meme, and concept. (The history of the terminology is discussed more fully in Lumsden and Wilson, 1981a.) This discussion has been further stimulated by the findings of Shannon (1948) and later students of information theory (Goldman, 1953; Brillouin, 1962; Khinchin, 1957; Gatlin, 1972; Chaitin, 1975), in which systems of transmissible information are objectified and ultimately ordered into quantifiable patterns composed of basic units.

Naming and defining an ultimate unit of culture has become a concern for gene-culture theorists as well as the cultural anthropologists. However, the nomenclatural diversity that exists in this field reflects the exploratory nature of the research more than basic disagreements. Despite persisting differences over detail and a minority opinion (e.g., Hallpike, 1982) that culture and its transmission cannot be resolved into basic units, the growing consensus is that the concept of units of culture makes sense. For example the term *culturgen* (L. *cultur(a)*, culture + L. *gen(o)*, produce) has been proposed in earlier work from our laboratory in order to attempt to specify a unit operationally: the information pattern corresponding to a relatively homogenous set of artifacts, behaviors, or mentifacts (mental constructions having little or no direct correspondence with reality), which either share without exception one or more attribute states used during psychological development, or at least share a consistently recurrent range of such attributes within a given polythetic set (Clarke, 1978; Lumsden and Wilson, 1980a, 1981a). As in anthropological definitions of culture, it is convenient to let such terms refer to the actual lifeways as well as to the corresponding information patterns when the difference is clear from context. Although somewhat formal, and by no means fully satisfactory, this initial conception of the culturgen allows the construction of sets of similar entities based on their role in cultural transmission. It also suggests empirical procedures for identifying and characterizing the units (Clarke, 1978; Lumsden and Wilson, 1981a). In assessing the merits of theories incorporating the concept of cultural units we have found it useful to note that the identification does not imply that culture is lacking holistic structure, any more than the specification of phonemes implies that human discourse is a set of unrelated elements.

Epigenesis is a term that refers to the total process of interaction between genes and the environment during development. During mental epigenesis, the substance of culture is reshaped to build the mind and its contents. Information encoded in the genes guides and shapes this development. The logic of the shaping is often expressible in the form of rules constraining various options or alternative pathways in the growth and differentiation of mental structures (Simon, 1979; Wexler and Culicover, 1980). It has been proposed that the term *epigenetic rules* be used to refer to these interacting components of the developmental logic (Lumsden and Wilson, 1981a). Contact with evolutionary genetics is made by admitting the theoretical possibility that changes in a gene can alter one or more epigenetic rules and the relations among them. In physiological terms, epigenetic rules of cognitive and behavioral development comprise one or more elements in a complex sequence of events occurring at various sites in the nervous system. Our analysis of the available data (Lumsden and Wilson, 1981a, 1982, 1983; Lumsden 1984a) has led us to conclude that these elements are crudely but usefully separated into two main categories: primary epigenetic rules, which regulate the development of systems ranging from the peripheral sensory filters to perception, and secondary epigenetic rules, which assemble the inner mental processes (Fodor, 1983), including the procedures of consciously deliberated valuation and decision making.

The epigenetic rules embody an innate part of the individual's strategy for learning culture. It is theoretically possible for this culture learning to belong to one of three classes of evolutionary strategies (Lumsden and Wilson, 1981a; Lumsden, 1983a, b, 1984a): pure genetic transmission, in which innate epigenetic rules prescribe essentially one developmental response to any culturgen or array of culturgens (hence, an entirely genetic culture is a theoretical possibility); pure cultural transmission, in which the epigenetic rules prescribe genetically unbiased use of any culturgen in competition with others in shaping mental development (this is the traditional viewpoint of cultural determinism in the social sciences; e.g., Harris, 1968; Freeman,

1983 for review); and gene-culture transmission, in which the innate epigenetic rules discriminate among multiple culturgens and are more likely to use some rather than others during their assembly of the mind. The term "gene-culture" in this context is not meant to reiterate the truism that both genes and culture somehow influence human development. Rather, it describes transmission in the presence of epigenetic rules that predispose mental development to take certain directions.

On the basis of present evidence it appears that much of human culture is sustained by gene-culture transmission rather than by pure cultural transmission. Whenever detailed studies have been conducted of development as mediated by choice among or directedness toward empirically distinguishable culturgens, they have almost always revealed an innate bias favoring some culturgens over others. Examples include a neonate preference for sugar combined with an active aversion to salty and bitter flavors (Maller and Desor, 1974; Chiva, 1979), affecting the evolution of cuisine; the innate discrimination for four basic hues (blue, green, yellow, red) (Bornstein *et al.*, 1976) and a greater ease of learning color classifications clustered on these color modes (Rosch, 1973), affecting the evolution of color-term systems (Berlin and Kay, 1969); infant phoneme discrimination, affecting later speech structure (Eilers *et al.*, 1977); infant preference for certain kinds of visual patterns (Fantz *et al.*, 1975; Hershenson *et al.*, 1965), regulating attention and arousal; neonate preference for normally composed facial features (a bias manifested within ten minutes following birth) (Freedman, 1974), and locomotor patterns (Fox and McDaniel, 1982), orienting the learner toward human sources of information; smiling and other specific forms of nonverbal communication (Eibl-Eibesfeldt, 1979), facilitating the development of bonding, reciprocity, and communication; nonverbal signals used in mother-infant bonding, inducing long-lasting affects in later maternal care (Klaus *et al.*, 1972; De Casper and Fifer, 1980); sexual differences in the carrying of infants and other larger objects (Salk, 1973; Lockhard, Daley and Gunderson, 1979); the fear-of-stranger response (Morgan and Ricciuti, 1973); the predisposition to acquire phobias against certain dangerous objects, such as heights, running water, and snakes, but not other dangerous objects, including electric sockets and guns (Marks, 1969); the development of sexual preferences within the family (Shepher, 1971; Wolf, 1968; Wolf and Huang, 1980; van den Berghe, 1983) affecting adult mating behavior and social structure; the size and operating speeds of long-term memory and short-term memory, affecting the choice of strategies in conscious deliberation and problem-solving (Simon, 1979); the development of linguistic knowledge (Chomsky, 1980; Wexler and Culicover, 1980), ontological knowledge, and knowledge about numerosity (Keil, 1981).

That the epigenetic rules have a genetic basis is indicated by several lines of evidence. Some epigenetic rules appear during early childhood and are relatively inflexible. In addition, pedigree analysis and standard comparisons of fraternal and identical twins, in some instances strengthened by longitudinal studies of development, have yielded evidence of genetic variance in virtually every category of cognition and behavior investigated by these means, including some that either constitute epigenetic rules or share components with them. These categories include color vision, hearing acuity, odor and taste discrimination, number ability, word fluency, spatial ability, memory, timing of language acquisition, spelling, sentence construction, perceptual skills, psychomotor skill, extroversion/introversion, homosexuality, proneness to alcoholism, age of first sexual activity, timing of Piagetian developmental stages, some phobias, certain forms of neuroses and psychoses, and others (for reviews see McClearn and Defries, 1973; Loehlin and Nichols, 1976; R. S. Wilson, 1978; Lumsden and Wilson, 1981a, 1983). Single gene variants have been identified that affect certain cognitive abilities selectively (Ashton *et al.*, 1979), as well as the ability to discriminate certain odorants (Amoore, 1977). It has also become apparent that mutations at a single locus can result in profound but highly specific changes in the architecture and operation of brain tissues such as mammalian neocortex (Caviness and Rakic, 1978; Rakic, 1979). Not only do these alterations modify behavior at the locomotory and perception levels, but they also introduce changes into such higher level functions as choice and decision (e.g., Bliss and Errington, 1977).

COEVOLUTIONARY MECHANISMS

Modern theories of gene-culture interaction can be sorted on the basis of degree to which the intervening mechanisms are described. Some theories (e.g., Feldman and Cavalli-Sforza, 1976; Rice *et al.*, 1978; Cavalli-Sforza and Feldman, 1981) use techniques of statistical correlation to symbolize the existence of a linkage without probing its workings. The "black box" of psychology, mental development, and behavior is not opened. Others (e.g., Boyd and Richerson, 1976; Richerson and Boyd, 1978; Pulliam and Dunford, 1980; Fagen, 1981) apply classical behavioral ecology in an effort to link genes directly to external behaviors. It is our view that evolutionary theories such as these, which do not explicitly deal with mentation and mental development, are not fully adequate descriptions of the human case. For humankind, genetic determinism shapes (at most) the epigenetic rules of mental development, not finished thoughts or behaviors. These are emergent properties of the growth process and rely on cultural input into the epigenetic rule systems.

We have recently investigated the feasibility of formal theories based on this perspective (e.g., Lumsden and Wilson, 1981a, 1983; Lumsden 1983a, 1984; Gushurst and Lumsden, 1984), with encouraging preliminary results. Our findings extend and in some cases revise inferences obtained from earlier methods of behavioral genetics and sociobiology. Comparison of theoretical approaches (e.g., Lumsden and Wilson, 1981a) indicates that between genes and culture two particular intermediate levels of organization are especially pertinent to the correct formal delineation of the linkage. They are also especially accessible in terms of available empirical data and theoretical representation (Figure 1). The first is cellular development in the brain, leading to the formation of nerve cell circuits. The second is cognitive development, in which the activity of the epigenetic rules builds mental structures and the capacity for overt behavior.

Acts of overt behavior in particular settings are the beginnings of large-scale social phenomena. Culture and macrosocial patterns in turn form a principal part of the environment in which human genetic evolution takes place (Holloway, 1966; Tobias, 1981; Lumsden, 1983b). The genetic fitness of individuals is determined by the impact of their choices on survival and reproduction. To take a hypothetical example pertinent to the sociobiological controversy of the 1970's, we note that a social deviant, physically fit though she or he may be, may find it hard to mate. A devout Marxist in Washington (circa 1950) would have been culturally selected against regardless of that individual's physical ability to survive, and thus "Marxist" genes (of classical sociobiological conception) would not be favored (*pace* Lewontin, 1981). The relation between individual genetic fitness and choice of deployed behaviors has in fact been explicitly documented in a wide array of real behavioral categories, including diet (Gajdusek, 1970), body marking (Blumberg and Hesser, 1975), sexual conventions (Daly and Wilson, 1978), marital customs (Daly and Wilson, 1978), economic practice (Irons, 1979), and others. The fitness of an individual is the product of survivorship and age-specific fertility integrated through the life cycle. Differential genetic fitness, resulting in the natural selection of competing genotypes, appears to be an important factor in the shaping of the gene ensembles that prescribe epigenetic rules (Lumsden and Wilson, 1981a; Lumsden 1983a, b, c; 1984a).

The transiton from genes to culture can therefore be represented in

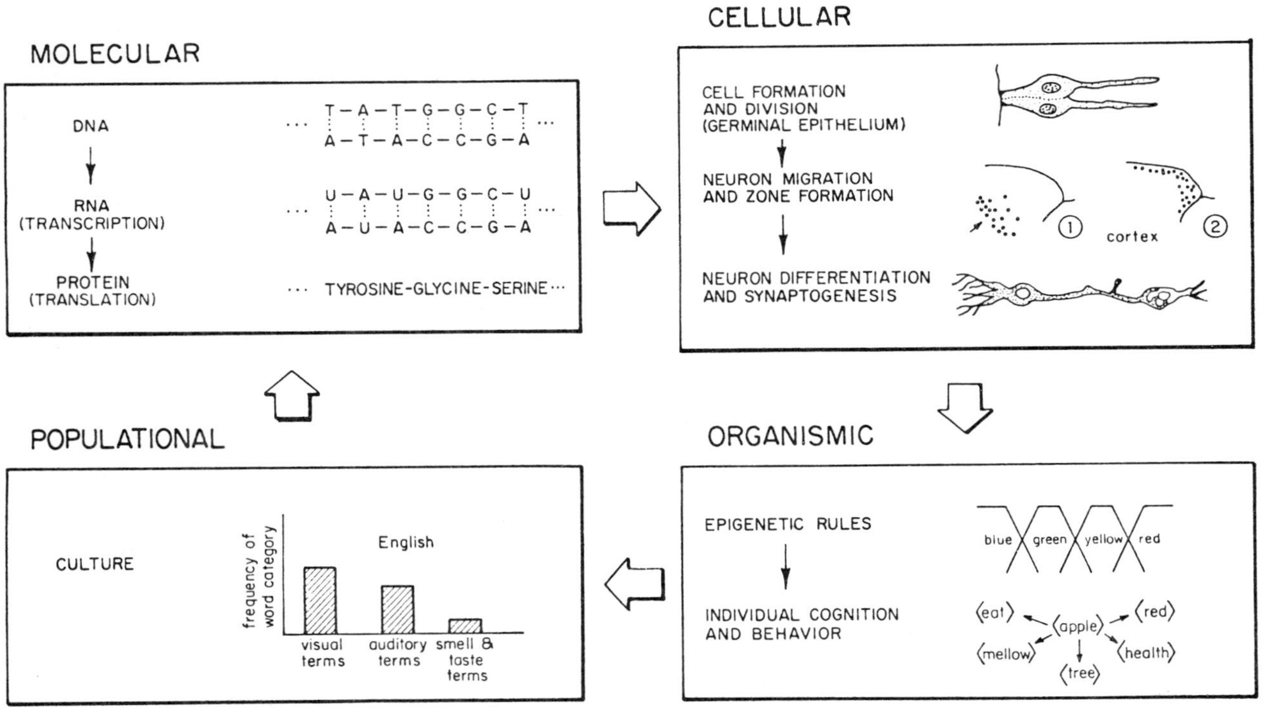

Fig. 1. An early depiction of the full circuit of causation in gene-culture coevolution, illustrating the grouping of domains into four principal levels of organization. Epigenesis takes place on the molecular, cellular, and organismic levels. Cultural events feed back to affect gene frequencies through natural selection and other evolutionary effects. The structure of the levels in the circuit, and the relationships among the levels, are now understood in greater detail than previously. (From Lumsden and Wilson, 1981a, reproduced with permission.)

initial approximation by a progression around a circuit of interactions rather than by steps up a ladder of independent levels of organization. The feedforward from genes to culture, via individual development and behavior, combines with the feedback from culture to genes through evolutionary pressures to link biological phenomena to social events. One of the advantages of such an approach to gene-culture interaction is its avoidance of a reductionistic strategy in favor of a formal treatment of this stepwise emergence from genes through the mind to culture. These procedures may be of relevance to Marxian social theorists and other students of large-scale social forms, who often stress the holistic, emergent nature of social phenomena (see, e.g., the citations and discussion in Lumsden and Wilson, 1981b).

Of course, pertinent theories of gene-culture coevolution must be able to solve problems that social scientists find interesting. In this respect, we have been struck by the lack of progress achieved with "general systems thinking" (Bertalanffy, 1968; Laszlo, 1972) in the social sciences. Statements that societies are complex assemblages of interacting parts, with capacities for self-regulation or autonomous change, are useful as a beginning. However, they do not appear to have produced theories that can consistently generate significant models of history and social structure. We have concluded that a segment essential to achieving this type of epistemological adequacy, once the role of the epigenetic rules has been clarified, is the transition from the individual to the society as a whole. This step in the coevolutionary circuit crosses the seam between biology and the social sciences. The resulting task of theory construction is challenging, and only partially completed at the present time. To describe social as well as biological events one must account for the correlated activity of many individuals in the presence of multiple culturgens, for their social patterns, and for the feedback of these patterns to cognition. In evolutionary time, on the scale of social history and genetic change, multiple genotypes with different epigenetic rules and cognitive properties interact with shifting cultures. The effects of fitness and other evolutionary forces can be obtained if life histories embodying predictions about individual survival and reproduction are evaluated for each genotype in the interacting assembly.

By combining methods from social theory with techniques from population genetics and the study of collective phenomena, it has been possible to formalize such gene-culture theories for both simpler (e.g., egalitarian hunter-gatherer) and more complex (e.g., stratified horticulatural) societies (Lumsden and Wilson, 1980a, 1981a; Lumsden 1983b, c, 1984a; Cavalli-Sforza and Feldman, 1981). Modeling the full gene-culture circuit in the sense of Figure 1 has been of particular interest in our own work. The procedure we have followed is to recognize the inherently probabilistic nature of thought and history. We thus deal with ensembles or sets of cultures rather than single instances. In the simplest cases of relevance to the social sciences, balance equations are used to specify the (time-dependent) probability distributions of social forms in relation to the thoughts and activities of individuals. Such probability distributions, or *ethnographic curves* (Lumsden and Wilson, 1980a, 1981a), predict the patterns of culture and social structure in the form of histograms of occurrence frequency, of the kind commonly calculated by cultural anthropologists during cross-cultural statistical surveys. Ethnographic curves are emergent phenomena characteristic of societies considered as integrated wholes. The mechanism sustaining their emergence is the biological feedforward from the genes through the innate epigenetic rules of cognitive development to the formation of social patterns.

The best-studied formal models based on these methodological principles are still relatively simple: for example, coevolution in an egalitarian society in which an allele segregating at a single genetic locus affects an epigenetic rule that regulates a bifurcation or "binary choice" in the pathway of mental development or decision-making (Lumsden and Wilson, 1981a; Lumsden, 1983a, 1984a). In judging the relevance of such idealizations, it is important to recall that the vast proportion of hominid evolution took place within hunter-gatherer bands that may have been sites of near-egalitarian social interaction (Lee and DeVore, 1976). Moreover, major genes at single loci are now known to exert effects selectively on the mode of operation of individual epigenetic rules (e.g., Kovach, 1980) and on the structure of brain circuitry (Rakic, 1979). Furthermore, deliberation of binary alternatives is a primary characteristic of choice and decision-making in everyday thinking (Levi-Strauss, 1966; Hallpike, 1979), and may be a major developmental strategy (Oster and Wilson, 1976).

A number of general findings have emerged from studies of theoretical models involving the entire coevolutionary circuit (Lumsden and Wilson, 1980a, b, 1981a, 1983; Lumsden 1983a, b, c, 1984a, b). The first is that even small innate biases in the epigenetic rules can be amplified in the resulting ethnographic curves. Absolute differences as low as 0.02 in the probabilities of choice among competing culturgens and modes of behavior can, under conditions of close social living, generate differences 1.5 times or more greater in the corresponding modes of the ethnographic curves. A series of amplification laws has been reported that provides quantitative relations between genertic changes in the epigenetic rules and the resulting shifts in the form of the ethnographic curves (Lumsden and Wilson, 1981a). Analysis of the available data on infant development (e.g., Maller and Desor, 1974; Freedman, 1974; Lockhard *et al.*, 1979) suggests that human epigenetic rules are in fact strong enough to create this type of effect. These conclusions support recent and independent results from social theory that micromotives can and do have a surprisingly large impact on macrobehavior (Schelling, 1978), and suggest that under the conditions typical of human societies, the feedforward form genes through the mind to culture is remarkably strong.

A second finding is the complement of the first: even when the options allowed by the epigenetic rules are rigidly constrained by genetic prescription, they can still generate wide cultural diversity. This variation is further enhanced by the culture-dependence of the effects produced by the epigenetic rules and is reinforced by the probabilistic nature of individual thought and behavior. Likely forms of ethnographic curves along with quantification of the gene-culture linkage have been obtained in a number of instances, including brother-sister incest avoidance, Amerindian village splitting, and, at the level of societal superstructure, fashion aesthetics in Western societies (e.g., Lumsden and Wilson, 1980b, 1981a, 1983; Lumsden 1983a, 1984a). In cases tied quite directly to survival and reproduction, such as incest, color vision, and certain instances of agricultural production, it has been possible to estimate the strength of the effects that genetic and cultural evolution exert on one another (e.g., Lumsden and Wilson, 1981a, 1983; Durham, 1982).

A third result is that the *tabula rasa* state of innately undirected mental development, leading to cultural determination of individual thought and behavior, is generally found to be unstable. Under conditions that entail differential fitness and no bifurcations to allelic polymorphism under frequency-dependent selection (see below), genotypes prescribing a perfect lack of bias toward the Darwinian-advantageous culturgens are rapidly replaced by genotypes prescribing a favourable bias and the maintenance of culture via gene-culture transmission. This result is consistent with the directedness observed to date in all classes of human epigenetic rules known to us. Although cultural evolution toward civilizations and other forms of complex societal organization (e.g., Lenski and Lenski, 1970) was previously thought to weaken the effects of biological evolution, this may not be the case. When the effects of a number of properties characteristic of later human social orders, such as stratification and increasing cultural complexity, are considered, they are found to favor the increase rather than the decrease of the number and complexity of innate epigenetic rules coding for dual inheritance via gene-culture transmission (Lumsden, 1983a).

A fourth finding is that the genetic evolution of human nature can proceed much more quickly than implied in most earlier discussions (Lumsden and Wilson, 1981a, 1983). Even when selectional forces are only moderate and the innate directedness provided by the epigenetic rules is mild in comparison with that generally observed in human developmental studies, the rate of change of gene frequency can be high enough to achieve the near-replacement of one allele by another within as few as 20 to 50 generations, or about 1000 years of human history. Genetic change via gene-culture coevolution is especially likely to have taken place during periods of time in which relatively little social change occurred and culture was especially easy to track genetically. The Paleolithic interval contains most of the biological history of the genus *Homo*, and during this time the rates of innovation and basic changes in the social system were relatively slow (Isaac, 1972). It therefore appears likely that opportunities have existed through prehistoric and historical times for genes to track the most successful forms of culturally transmitted knowledge and behavior. Empirical examples of this tracking process have recently been proposed (Lumsden, 1983a, b).

Finally, because people use the decisions and behaviors of others as part of their own deliberations, natural selection of genotypes during gene-culture coevolution is frequency dependent (Lumsden and Wilson, 1981a, b; Lewontin, 1981). Its direction and intensity are determined in part by the number of individuals using alternative culturgens and thus by the relative frequencies of the competing gene variants. Classical sociobiological arguments (e.g, Wilson, 1975, 1978; Dawkins, 1976; Barash, 1978; Alexander, 1979; Barlow and Silverberg, 1980) have usually assumed that some measure of fitness (such as inclusive fitness) is optimized during human evolution. However, if this finding of gene-culture theory turns out to have general validity, the basic assumption of ordinary fitness optimization and the many models based on it become suspect. Frequency-dependent selection may not lead to the optimization of a fitness measure at all. The pathways of both genetic and cultural evolution may exhibit behavior much more complex than an approach to a coevolutionarily stable optimum. A diversity of genetic and culturgenic variants can persist through time, even changing from generation to generation in a truly chaotic fashion (Lumsden and Wilson, 1981a: pp. 298–302). For such systems arguments from optimization are wrong and must be replaced by methods using the first principles of coevolutionary genetics.

ETHICAL EPISTEMOLOGY: COEVOLUTION AND THE CULTURAL SUPERSTRUCTURE

Gene-culture theories have been shown to provide new insights concerning basic elements of adaptive cultural processes, such as incest avoidance, conflict and societal growth, disease and its cultural treatment, and cultural complexity. What, however, can be said about characteristics tied less directly to matters of survival and reproduction, and often described (in the Marxian sense) as elements of a culture's "superstructure": literary discourse, aesthetic achievement, systems of ethical and political reasoning, and so on? Applications of coevolutionary theory have recently been made to the superstructural problems of linguistic color-term systems and aesthetic reflected in female bodily adornment (Lumsden and Wilson, 1981a). However, the studies to date have been preliminary and the feasibility of highly effective superstructural applications has received relatively little attention. We would like to suggest that formal theories of gene-culture coevolution will be highly productive in such applications. By design they synthesize the core elements of population genetics, neurobiology, cognitive science, and social theory required for a coherent attack on advanced mental phenomena, including those involved in the production of superstructural culturgens. Moreover, because culture and mind can now be dealt with more explicitly within formal evolutionary theory, the results are likely to produce novel additions to the results of earlier evolutionary reasoning. Since they combine all the complexities of human consciousness, free will, and social convention, the subjects of ethics and moral philosophy have often been considered domains separated in principle from the sciences. The prospect of connecting them with evolutionary biology through the creation of a deep bioethics therefore provides a useful illustration of our main point about the promise of coevolutionary theories in superstructural applications.

To the sociobiologist, a striking feature of ethics is its fundamental role in the structure of human societies. Human reality is a moral reality. Details as to what, specifically, is "right" or "wrong" vary from society to society, but the obsession to regulate human behavior using ethical principles is universal. Indeed, our Western heritage is dominated by philosophers, theologians, and poets who invariably discuss the nature of value and the quest for virtue in human life.

The learning of ethical principles involves both the assimilation of culturally transmitted information and the expression of individual preferences. Ethical belief systems are encoded, and maintained culturally, but the individual must still choose a personal stand on any given issue. Because of this there is ethical variation within a population. Specific values may change from individual to individual or even generation to generation. However, while ethical systems can shift in emphasis, they never disappear. A younger generation might reject much of the "morality" of its parents' generation, but will nonetheless replace it with its own codes of ethics.

It is these characteristics of ethics – its universal importance in both social and mental life, the fact that ethical behavior is replace with consicious deliberated choice and decision, and the fact that ethical systems are rooted both in reason and passion – that the sociobiologist must explain. With the publication of *Sociobiology: The New Synthesis*, in 1975, E. O. Wilson started yet another round in the long debate over human ethical systems. The controversy generated by Wilson's sociobiological analysis focussed on two issues. One was the scientific interpretation of concepts fundamental to "being human": for example, self-determinism, consciousness, and free will. This was combined with a fear of the possible consequences sociobiology as an advocation of Social Darwinism and as the basest of utilitarian ethics. Concern was also expressed by those who, while intrigued with the application of sociobiology to human behavior, found serious gaps in the classical sociobiological explanations of such aspects of human behavior as conscious rational thought and individual moral development. (For a review of the history of these controversies, see Lumsden and Wilson, 1983.)

Classical sociobiological models (Wilson, 1975, 1978; Singer, 1981; Murphy, 1982; Maxwell, 1984) tend to equate the human moral passions to reactions that are hardwired into the brain's limbic systems. They claim that the associated behaviors have been evolutionarily fixed as adaptive strategies, with attached feelings that are automatically called out of the brain. For example, incest avoidance is a good strategy: inbreeding leads to deformities and reduces variation. Thus, incest avoidance behavior, and the deep inner feelings that incest is wrong, are evolutionarily strategic. Far from being evil or virtuous, incest is simply unadvantageous. Thus human judgements of right and wrong are based on a very relative ethics determined by what optimizes the appropriate Darwinian fitness in particular circumstances. Rationality, reason, and conscious reflection serve, but do not establish, the motivational goals wired into the limbic system. Human ethics, according to this model, is an ethics in which value is ultimately an expression of reproductive and survival advantage.

Gene-culture theories and their supporting data indicate that the complexity of human ethical behavior is not reducible to genes and hardwired responses. Philosophical problems aside, concepts of hardwired feelings and reactions leave scant room for conscious human choice (Lumsden and Wilson, 1983) or free will or the cultural transmission of ethical beliefs. The actual learning of ethical principles has been studied to some extent. Kohlberg (1969, 1976; see also Turiel, 1978) and Gilligan (1982) have identified stages this process moves through in the male and female psyche. According to these analyses, there is a very set way in which human children learn ethical systems, and this process can be identified in any child. While every child is capable of learning ethical systems, and even seems predisposed to so so, what each learns varies from indivudla to individual and culture to culture. However, there are certain common features that can be identified in the ethical learning process, and in the content of the final belief systems. Gene-culture theories (Lumsden and Wilson, 1983) together with developmental data (review in Keil, 1981), point to the existence of epigenetic rules that make the culture-learner more likely to acquire certain systems of ethical knowledge rather than others. So, for example, there are no innate feelings of moral revulsion about incest; rather there are epigenetic rules, making it likely that judgements of evil are readily attached during mental development to human activities involving incestuous behavior.

In ethical situations, the individual will often feel strongly that one or another course of action is the "right" thing to do. However, there is generally also an accompanying rationalization or (in especially fortuitous circumstances) logical justification of this choice. This is of particular importance in a social context, where the individual must explain to others the reasons behind his or her choice of action. Social approval is important to individual adaptation in human societies (Lumsden and Wilson, 1981a, pp. 297–298), and it is essential to the development of moral knowledge (e.g., the "good-boy" stage in the Kohlberg sequence, Kohlberg, 1969). It is rare that something is accepted as right or wrong just because "it is" (i.e., the number of self-evident propositions underlying an ethical system is usually rather small in proportion to the known derived arguments). This type of "just so" argument is unsatisfactory for most people, and may be much harder to learn.

The ethical center in human social life can therefore be expected to create selection pressures favoring certain forms of the moral passions. The associated genetic basis will in turn influence the sorts of ethical knowledge assimilated during enculturation. A culturally fit ethical system is one that can be reliably transmitted and upheld. If a system existed in which values or actions could not be verbally defended, that set of beliefs would be open to questioning and alteration. Coevolutionarily successful moral passions will therefore tend to have self-evident propositions as culturgenic counterparts. These propositions are readily attached to lore of rationalization through which value is interpreted and persuasively justified. Given the expansive tendencies of human reason (cf. Singer, 1981), the consistency and covering power of these propositions under logical derivation and (possibly) generalization will also test their cultural fitness in competition with alternative ethics. Parts of the system found to be inconsistent with the basis will be vulnerable to change. Thus any coevolutionarily successful ethical system is expected to be logical, rationalizable, and (with respect to moral sensibilities) appealing in the context of everyday thought. Naturally, systems that inherently favor one segment of a society over another will not be stable if those oppressed gain political or socio-economic power and tnen refuse to accept the ethical *status quo*. Sometimes, as with women's rights in the United States, the changes are slow. But instances of rapid change, such as the liberalization of Western sexual mores in the sixties and seventies, occur as well. Often in such cases, change does not occur until those affected by that part of the ethical system begin to reject it.

Classical sociobiology is often interpreted as predicting a graudal spread of utilitarian ethics at the expense of other, possibly more humane, systems of moral conduct, although it is by no means evident that this interpretation is entirely sound (see e.g., the reference to deontology in Wilson, 1975, p. 563). The obvious criticism to this stand is that moral consciousness in real human beings is not completely utilitarian (Beauchamp, 1982). The conventional expectation may be based on an unwarranted generalization: the fact that an ethical system optimizes evolutionary utility does not necessarily imply that the system is itself one of utilitarianism. The problem, in fact, is open: does gene-culture coevolution of the kind seen in human populations favour utilitarian ethical systems, or systems of some other form?

It seems to us that the coevolutionarily stable system is likely to be deontological, or at least contain a strong component of these beliefs. In contrast with deontological reasoning, choosing a best utilitarian course of action requires that an outcome of greatest value be perceived – a difficult problem in optimization given the limited power (Simon, 1979) of human cognition. Moreover, a social group unaware of its epigenetic rules can experience moral intuitions but cannot effectively interpret where they come from. Shaped by epigenetic rules for defensible moral passions and self-evident propositions, it continues to live by the "conscience" of its members and, under rationalization, by "God's will", dutifully expressed in the details of individual choice and action (Lumsden and Wilson, 1981; Gushurst and Lumsden, 1984). Unable to foresee all significant ends, it members feel the necessity to follow their "consciences", even though the consequences might not be favorable at all times. The ideas of duty and virtue become the defensible cultural innovations that compensate for less than optimal ends and make means a strategy of adaptive value. Culture, dependent upon a smoothly functioning society, readily selects for individual predispositions for acquiring notions of duty and conscience, capitalizing on the susceptibility of a smart primate to the nuances of reciprocal altruism and social obligation.

The data available to us from developmental psychology appear to be in general accord with these expectations. In cross-cultural and longitudinal studies of moral judgement, sometimes extending over two decades, Kohlberg and his associates (e.g., Kohlberg, 1969; Colby *et al.*, 1983) found that the dominant mode of reasoning used by adults in their samples was that of their "Stage 4". Although the definition of Stage 4 has changed somewhat over the years, both the original and contemporary descriptions depict a cognitive style rich in deontological content: "Stage 4: Authority and social order-maintaining orientation. Orientation to 'doing duty' and to showing respect for authority and maintaining the given social order for its own sake. Regard for earned expectations of others." (Kohlberg, 1969, p. 376). In later work, what is right amounts to "Fulfilling the actual duties to which you have agreed. Laws are to be unheld in extreme cases where they conflict with other fixed social duties. Right is also contributing to society, the group, or institution." The reasons for doing right are "To keep the institution going as a whole, to avoid the breakdown in the system if 'everyone did it', or the imperative of conscience to meet one's defined obligations." (Colby *et al.*, 1983). Moreover, Stage 3, which precedes Stage 4 in Kohlberg's proposed developmental sequence, and whose content is often present alongside Stage 4 in an adult's moral reasoning, emphasizes the Golden Rule and its general application to judgement situations (Colby *et al.*, 1983, pp. 44–56). There may be some interesting patterns of sex difference in this trends. Whereas Kohlberg's work emphasizes the moral judgements of males, Gilligan (1982) has begun to redress the balance with studies of women's moral reasoning. Although her assessment strategy is more anecdotal and less longitudinal than the methods used in the Kohlberg studies, her interviews indicate a strong ends-orientation among her participants along with a distinctive connection of morality to motions of responsibility in relationships (Gilligan, 1982; see also Saxton, 1981, p. 66). In the absence of more detailed information it is, however, too early to state whether the mix of deontological content in women's moral thought is significantly different from that in men.

Relative to sanctioned behavior, most civilizations and societies have ethical principles in common that would be predicted using sociobiological ideas that are by now familiar: prohibition of wanton killing, incest avoidance, protection of children, preferential assistance to kin, allegiance to higher, often abstract, authority, etc. These principles are then stratified into a system of rules and practices that differ from society to society. Any explanation of ethical systems is just that, of course, and is not to be mistaken (here or elsewhere) for advocation of any particular ethical reality. In terms of prescription, it has been suggested (Nozick, 1981) that human evolution is trying to track an objectively existing ethical system, which if used would optimize our fitness. To date, however, the problems of supplying existence proofs for the objective system, the tracking process, and the associated coevolutionary optimum are still open. An objective ethical reality that is evolutionarily reachable therefore remains an unproven, but tantalizing, possibility.

CONCLUSIONS

Some social scientists have suggested that an explicit concern for biology in studies of culture is unnecessary (Sahlins, 1976; Caplan, 1978; Harris, 1979; Bock, 1980; Montagu, 1980; Trigg, 1982; Marks *et al.*, 1983). Biological underpinnings or even constraints may exist, but they are common from culture to culture and form a universal raw material that culture proceeds to shape. The differences among cultures, together with the essential nature of social diversity, can be understood by reference to sociocultural patterns and processes, taking biology as a common denominator among humans. Why one culture differs from another is thus a question that does not concern biology (e.g., Goodenough, 1983; for a history of this position together with an opposing view by a cultural anthropologist, see Freeman, 1983).

The evidence summarized in this report indicates that views such as these are wrong. In the presence of gene-culture linkages sustained by epigenetic rules, "explanations" of cultural differences or social diversity by reference only to the variations between information contents of societies are not explanations at all. They are only recitations of the boundary conditions under which these human differences developed. Without simultaneous reference to genes, mind, and culture, one is at a loss to explain why certain environmental or sociocultural events led to their specific consequences for

Fagen, R.: 1981, *Animal Play Behavior*, Oxford University Press, New York.

Fantz, R. L., Fagan III, J. F., and Miranda, S. B.: 1975, Early Visual Selectivity: As a Function of Pattern Variables, Previous Exposure, Age from Birth and Conception, and Expected Cognitive Deficit, in L. B. Cohen and P. Salapatek (ed.), *Infant Perception: From Sensation to Cognition*. Vol. 1: *Basic Visual Processes*, Academic Press, New York, pp. 249–345.

Feldman, M. W. and Cavalli-Sforza, L. L.: 1976, 'Cultural and Biological Evolutionary Processes, Selection for a Trait under Complex Transmission', *Theoretical Population Biology* 9, 238–259.

Fodor, J. A.: 1983, *The Modularity of Mind: An Essay on Faculty Psychology*, MIT Press, Cambridge, MA.

Fox, R. and McDaniel, C.: 1982, 'The Perception of Biological Motion by Human Infants', *Science* **218**, 486–487.

Freedman, D. G.: 1974, *Human Infancy: An Evolutionary Perspective*, Lawrence Erlbaum Association, Hillsdale, NJ.

Freeman, D.: 1983, *Margaret Mead and Samoa: The Making and Unmaking of an Anthropological Myth*, Harvard University Press, Cambridge, MA.

Gajdusek, D. C.: 1970, 'Physiological and Psychological Characteristics of Stone Age Man', *Science and Technology* **33**, 26–62.

Garn, S. M.: 1964, *Culture and the Direction of Human Evolution*, Wayne State University Press, Detroit.

Gatlin, L. L.: 1972, *Information Theory and the Living System*, Columbia University Press, New York.

Gilligan, C.: 1982, *In a Different Voice: Psychological Theory and Women's Development*, Harvard University Press, Cambridge, MA.

Goldman, S.: 1953, *Information Theory*, Dover, New York.

Goodenough, W.: 1983, 'Margaret Mead and Cultural Anthropology', *Science* **220**, 906, 908.

Gould, S. J.: 1983, 'Sociobiology Goodbye', *The New York Review of Books* **30** (11): 5, 6, 8, 10.

Gushurst, A. C. and Lumsden, C. J.: 1984, 'The Morality of the Culturgen', *Zygon*, to be submitted.

Hallpike, C.: 1982, 'The "Culturgen": Science or Science Fiction', *Behavioral and Brain Sciences* **5**, 12–13.

Harris, M.: 1968, *The Rise of Anthropological Theory: A History of Theories of Culture*, Harper and Row, New York.

Harris, M.: 1979, *Cultural Materialism: The Struggle for a Science of Culture*, Random House, New York.

Hershenson, M., Munsinger, H., and Kessen, W.: 1965, 'Preference for Shapes of Intermediate Variability in the Newborn Human', *Science* **147**, 630–631.

Holloway, R. L.: 1966, 'Cranial Capacity, Neural Organization, and Hominid Evolution: A Search for More Suitable Parameters', *American Anthropologist* **68**, 103–121.

Hutchinson, A.: 1970, *Labanotation or Kinetography Laban: The System of Analyzing and Recording Movement*, Theatre Arts Books, New York.

Isaac, G. Ll.: 1972, 'The Chronology and Tempo of Cultural Change During the Pleistocene', in W. W. Bishop and J. A. Miller (eds.), *Calibrating Hominoid Evolution*, Scottish Academic Press, Edinburgh, pp. 381–430.

Irons, W.: 1979, 'Cultural and Biological Success', in N. A. Chagnon and W. Irons (eds.), *Evolutionary Biology and Human Social Behavior: An Anthropological Perspective*, Duxbury Press, North Scituate, MA, pp. 257–272.

Keesing, R. M.: 1976, *Cultural Anthropology: A Contemporary Perspective*, Holt, Rinehard and Winston, New York.

Keil, F. C.: 1981, 'Constraints on Knowledge and Cognitive Development', *Physiological Reviews* **88**, 197–227.

Khinchin, A. I.: 1957, *Mathematical Foundations of Information Theory*, transl. by R. A. Silverman and M. D. Friedman, Dover, New York.

Klaus, M. H., Jerauld, R., Kreger, N. C., McAlpine, W., Steffa, M., and Kennel, J. H.: 1972, 'Maternal Attachment: Importance of the First Post-partum Days', *New England Journal of Medicine* **286**, 460–463.

Kohlberg, L.: 1969, Stage and Sequence: The Cognitive-Developmental Approach to Socialization, in D. A. Goslin, (ed.), *Handbook of Socialization Theory and Research*, Rand McNally and Co., Chicago, pp. 347–480.

Kohlberg, L.: 1976, Moral Stages and Moralization: The Cognitive-Developmental Approach, in T. L. Lickona (ed.), *Moral Development and Behavior: Theory, Research and Social Issues*, Holt, Rinehart, and Winston, New York, pp. 31–53.

Kovach, J. K.: 1980, 'Mendelian Units of Inheritance Control Color Preference in Quail Chicks (*Coturnix coturnix japonica*)', *Science* **207**, 549–551.

Kroeber, A. L. and Kluckhohn, C.: 1963, *Culture: A Critical View of Concepts and Definitions*, Vintage Books, Random House, New York.

Laban, R.: 1975, *Laban's Principles of Dance and Movement Notation*, 2nd ed., MacDonald and Evans, London.

Laszlo, E.: 1972, *Introduction to Systems Philosophy: Toward a New Paradigm of Contemporary Thought*, Harper and Row, New York.

Lee, R. B. and DeVore, L. (eds.): 1976, *Kalahari Hunter-Gatherers: Studies of the !Kung San and Their Neighbors*, Harvard University Press, Cambridge, MA.

Lenski, G. and Lenski, J.: 1970, *Human Societies: A Macrolevel Introduction to Sociology*, McGraw-Hill Book Co., New York.

Levi-Strauss, C.: 1966, *The Savage Mind*, University of Chicago Press, Chicago.

Lewontin, R. C.: 1981, 'Sleight of Hand', *The Sciences* **21**, 23–26.

Lockhard, J. S., Daley, P. C., and Gunderson, V. M.: 1979, 'Maternal and Paternal Differences in Infant Carry: U.S. and African Data', *American Naturalist* **113**, 235–246.

Loehlin, J. C. and Nichols, R. C.: 1976, *Heredity, Environment, and Personality*, University of Texas Press, Austin, TX.

Lopreato, J.: 1984, *Human Nature and Biocultural Evolution*, Allen and Unwin, Boston.

Lumsden, C. J.: 1983a, 'Gene-Culture Coevolution and the Devolution of Tabula Rasa', *Jouranl of Social and Biological Structures* **6**, 101–114.

Lumsden, C. J.: 1983b, 'Neuronal Group Selection and the Evolution of Hominid Brain Size', *Journal of Human Evolution* **12**, 169–184.

Lumsden, C. J.: 1983c, 'Gene-Culture Linkages and the Developing Mind', in C. J. Brainerd (ed.), *Recent Advances in Cognitive-Developmental Theory*, Springer-Verlag, New York, pp. 123–166.

Lumsden, C. J.: 1984a, 'Parent-Offspring Conflict over the Transmission of Culture, *Behavioral Ecology and Sociobiology*, in press.

Lumsden, C. J.: 1984b, 'Sociobiology', in G. Adelman (ed.), *The Encyclopedia of Neuroscience*, Birkhauser Poston, Inc.

Lumsden, C. J.: 1984c, 'Dual Inheritance in Haploid Organisms: A Model of Magnetotactic Bacteria', *Journal of Theoretical Biology*, in press.

Lumsden, C. J. and Wilson, E. O.: 1980a, 'Translation of Epigenetic Rules of Individual Behavior Into Ethnographic Patterns', *Proceedings of the National Academy of Sciences of the United States of America* 77, 4382–4386.

Lumsden, C. J. and Wilson, E. O.: 1980b, 'Gene-Culture Translation in the Avoidance of Sibling Incest', *Proceedings of the National Academy of Sciences of the United States of America* 77, 6248–6250.

Lumsden, C. J. and Wilson, E. O.: 1981a, *Genes, Mind, and Culture: The Coevolutionary Process*, Harvard University Press, Cambridge, MA.

Lumsden, C. J. and Wilson, E. O.: 1981b, 'Genes, Mind, and Ideology', *The Sciences* **21**, 6–8.

Lumsden, C. J., and Wilson, E. O.: 1982, 'Mind and the Linkage Between Genes and Culture: A Precis of Genes, Mind, and Culture', *The Behavioral and Brain Sciences* **5**, 1–7.

Lumsden, C. J. and Wilson, E. O.: 1983, *Promethean Fire: Reflections on the Origin of Mind*, Harvard University Press, Cambridge, MA.

Maller, O. and Desor, J. A.: 1974, 'Effect of Taste on Ingestion by Human Newborns', in J. Bosma (ed.), *Fourth Symposium on Oral Sensation and Perception: Development in the Fetus and Infant*, Government Printing Office, Washington, DC, pp. 279–811.

Marks, I. M.: 1969, *Fears and Phobias*, Academic Press, New York.

Marks, J., Staski, E., and Schiffer, M. B.: 1983, 'A Return to Basics', *Nature* **302**, 15–16.

Maxwell, M.: 1984, *Human Evolution*, Columbia University Press, New York.

McClearn, G. E. and Defries, J. C.: 1973, *Introduction to Behavioral Genetics*, W. H. Freeman, San Francisco.

Medawar, P.: 1981, 'Stretch Genes', *New York Review of Books* **28** (12), 45, 47, 48.

Montagu, A. (ed.): 1980, *Sociobiology Examined*, Oxford University Press, New York.

Morgan, G. A. and Ricciuti, H. N.: 1973, 'Infants' Response to Strangers During the First Year', in L. J. Stone, H. T. Smith and L. B. Murphy (eds.), *The Competent Infant: Research and Commentary*, Basic Books, New York, pp. 1128–1138.

Murdock, G. P.: 1967, 'Ethnographic Atlas: A Summary', *Ethnology* **6**, 109–236.

Murphy, J. G.: 1982, *Evolution, Morality, and the Meaning of Life*, Rowman and Littlefield, Totawa, New Jersey.

Naroll, R. and Cohen, R. (eds.): 1970, *A Handbook of Method in Cultural Anthropology*, The Natural History Press, Garden City, NY.

Nozick, R.: 1981, *Philosophical Explanations*, Harvard University Press, Cambridge, MA.

Ortner, D. J.: 1983, *How Humans Adapt: A Biocultural Odyssey*, Smithsonian Institution Press, Washington, DC.

Oster, G. F. and Wilson, E. O.: 1976, *Caste and Ecology in the Social Insects*, Princeton University Press, Princeton, NJ.

Pulliam, H. R. and Dunford, C.: 1980, *Programmed to Learn: An Essay on the Evolution of Culture*, Columbia University Press, New York.

Rakic, P.: 1979, 'Genetic and Epigenetic Determinants of Local Neuronal Circuits in the Mammalian Central Nervious System', in F. O. Schmitt and F. G. Worden (eds.), *The Neurosciences: Fourth Study Program*, MIT Press, Cambridge, MA, pp. 109–127.

Rice, J., Cloninger, C. R., and Reich, T.: 1978, 'Multifactorial Inheritance with Cultural Transmission and Assortative Mating. I: Description and Basic Properties of the Unitary Models', *American Journal of Human Genetics* **30**, 618–643.

Richerson, P. J. and Boyd, R.: 1978, 'A Dual Inheritance Model of the Human Evolutionary Process. I: Basic Postulates and a Simple Model', *Journal of Social and Biological Structures* **1**, 127–154.

Rosch, E.: 1973, 'Natural Categories', *Cognitive Psychology* **4**, 328–350.

Sahlins, M.: 1976, *The Use and Abuse of Biology: An Anthropological Critique of Sociobiology*, University of Michigan Press, Ann Arbor, MI.

Salk, L.: 1973, 'The Role of the Heartbeat in the Relations Between Mother and Infant', *Scientific American* **228**, 24–29.

Saxton, M.: 1981, 'Are Women More Moral than Men? An Interview With Psychologist Carol Gilligan', *Ms*., December, 64–66.

Schelling, T. C.: 1978, *Micromotives and Macrobehavior*, W. W. Norton, New York.

Schneider, L. and Bonjean, C. M.: 1973, *The Idea of Culture in the Social Sciences*, Cambridge University Press, Cambridge.

Schwartz, B. M. and Ewald, R. H.: 1968, *Culture and Society*, The Ronald Press Company, New York.

Shannon, C. E.: 1948, 'The Mathematical Theory of Communication', *Bell System Technical Journal* **27**, 379–423, 623–656.

Shepher, J.: 1971, 'Mate Selection Among Second-Generation Kibbutz Adolescents and Adults: Incest Avoidance and Negative Imprinting', *Archives of Sexual Behavior* **1**, 293–307.

Singer, P.: 1981, *The Expanding Circle: Ethics and Sociobiology*, Farrar, Straus and Giroux, New York.

Simon, H. A.: 1979, *Models of Thought*, Yale University Press, New Haven, CN.

Swanson, C. P.: 1983, *Ever-Expanding Horizons: The Dual Informational Sources of Human Evolution*, University of Massachusetts Press, Amherst, MA.

Taylor, R. B.: 1973, *Introduction to Cultural Anthropology*, Allyn and Bacon, Boston.

Tinbergen, N.: 1972, 'Functional Ethology and the Human Sciences', *Proceedings of the Royal Society of London, Series B* **182**, 385–410.

Tobias, P. V.: 1981, *The Evolution of the Human Brain, Intellect and Spirit. 1st Abbie Memorial Lecture, University of Adelaide, South Australia, 12th October, 1979*. University of Adelaide Information Office, Adelaide.

Trigg, R.: 1982, *The Shaping of Man: Philosophical Aspects of Sociobiology*, Basil Blackwell, Oxford.

Turiel, E.: 1978, 'The Development of Moral Concepts', G. S. Stent (ed.), *Morality as a Biological Phenomenon*, Dahlem Konferenzen, Berlin, pp. 125–140.

van den Berghe, P.: 1983, 'Human Inbreeding Avoidance: Culture in Nature', *The Behavioural and Brain Sciences* **6**, 91–123.

Wexler, K. and Culicover, P. W.: 1980, *Formal Principles of Language Acquisition*, MIT Press, Cambridge, MA.

Wilson, E. O.: 1975, *Sociobiology: The New Synthesis*, The Belknap Press of Harvard University Press, Cambridge, MA.

Wilson, E. O.: 1978, *On Human Nature*, Harvard University Press, Cambridge, MA.

Wilson, E. O. and Lumsden, C. J.: 1984, 'D'ou vient l'esprit. Les nouvelles frontières de la connaissance', *Cahier Sciences du Figaro-Magazine*, pp. 72–86.

Wilson, R. S.: 1978, 'Synchronies in Mental Development: An Epigenetic Perspective', *Science* **202**, 939–948.

Wolf, A. P.: 1968, 'Adopt a Daughter-in-Law, Marry a Sister: A Chinese Solution to the Problem of the Incest Taboo', *American Anthropologist* **70**, 864–874.

Wolf, A. P. and Huang, C. S.: 1980, *Marriage and Adoption in China, 1845–1945*, Stanford University Press, Stanford, CA.

PART I

SOCIOBIOLOGICAL CONCEPTIONS

JOSEPH F. HANNA

SOCIOBIOLOGY AND THE INFORMATION METAPHOR

ABSTRACT. One focus of "the sociobiological debate" has been the issue of genetic determinism. I argue that the debate has suffered from an inadequate epistemological foundation for language invoking the *information metaphor*. Talk, which purports to be purely *descriptive* about genetic codes, blueprints, and programs actually acquires its meaning by extension from talk about human agency, involving such *prescriptive* notions as intention and conscious purpose. Following Nagel, I analyze the concept of a biological program in terms of the role of programs in goal-directed systems. The most significant implications of the analysis are that: (i) a program is a *relation* between a physical structure (DNA) and a goal-directed system which it "controls", so that the same program (DNA), embedded in different biological systems, can have different "meanings"; and (ii) programs are always "programs for stability", so that there are no "biological programs for change". This concept of a biological program is applied to questions of degrees of genetic determinism in order to shed some light on the "nature-nurture" controversy.

1. THE ISSUE OF GENETIC DETERMINISM

The heated debate over the scientific legitimacy of sociobiology has been fueled by a complex web of conflicting presuppositions, attitudes, and personalities: presuppositions about human nature and the nature of the human sciences, attitudes regarding the relative value of different forms of government and different modes of scientific inquiry, and personalities molded by different perceptions of the relative significance of individuals and groups. Clearly the conflict surrounding sociobiology cannot be encompassed by any single issue. Nonetheless, from a philosophical point of view, one central, unresolved issue that continues to fuel the debate is the status of claims regarding the extent of genetic determinism. In his popular treatise, *On Human Nature*, E. O. Wilson (1978, p. 19) states the thesis of genetic determinism in a most forceful and dramatic way: "The question of interest is no longer whether human social behavior is genetically determined; it is to what extent. The accumulated evidence for a large heredity component is more detailed and compelling than most persons, including even geneticists, realize. I will go further; it is already decisive." Its antithesis, which does not quite amount to a claim of cultural or environmental determinism, is stated

James H. Fetzer (ed.), Sociobiology and Epistemology, 31–55.

with equal force and drama by the amorphous "Sociobiology Study Group of Science for the People" (1976, p. 285):

> We can dispense with the direct evidence for a genetic basis of various human social forms in a single word, 'none'. The genetics of normal human behavior is in a rudimentary state because of the impossibility of reproducing particular human genotypes over and over, or of experimentally manipulating the environments of individuals or groups. There is no evidence that meets the elementary requirements of experimental design, that such traits as xenophobia, religion, ethics, social dominance, hierarchy formation, slavemaking, etc., are in any way coded specifically in the genes of human beings.

It is apparent from these passages that both of the central parties to "the sociobiology debate" have supposed that the issue of genetic determinism can, in principle, be resolved on the basis of appropriate empirical evidence. The champions of sociobiology have claimed that an abundance of empirical evidence supporting the thesis of genetic determinism is already available. The critics of sociobiology have disputed the relevance of the evidence cited, especially to the question of *human* social behavior, while suggesting that it is clear how one could, in principle if not in practice, go about collecting the relevant empirical evidence to decide the issue. The *critical* aim of the present paper is to demonstrate that both parties are mistaken in their views – that we neither have empirical evidence that decides the issue of genetic determinism, nor do we have a clear idea of the sort of evidnece that would, in principle, decide the issue. I argue that, in this respect, we do not have an adequate epistemology within which to frame the fundamental thesis of sociobiology. The *constructive* aim of the present paper is to recommend a more adequate epistemological foundation for claims about degrees of genetic determinism.

The thesis of genetic determinism has been formulated in a variety of ways; for example, sociobiologists speak of genes causing behavior, influencing behavior, controlling behavior, determining behavior, constraining behavior, etc. From an epistemological point of view, all of these locutions are highly problematic, and the problems which they raise are closely related. The present analysis focuses upon the term 'control', because the concept of control is central to the "information metaphor" (involving talk of programs, codes, blueprints, etc.) – a metaphor that continues to dominate the biological sciences as a whole and is the principal vehicle for discussions of genetic determinism.

As noted above, the controversy surrounding sociobiology *cannot* be treated as a straightforward conflict between the thesis of genetic determinism and the antithesis of environmental determinism. In the first place, all parties to the debate emphasize the fact that every behavioral trait is a product of (is controlled by) both genes and environment; the issue is the relative importance of each of these factors. Thus, if one considers a continuum extending from strict genetic determinism at one extreme to strict environmental determinism at the other, various participants could be ordered on that continuum, with the sociobiologists tending toward the genetic end and the critics tending toward the environmental. However, the situation is more complicated: the continuum of positions that I have in mind presupposes a principle of determinism, according to which behavioral phenotypes are completely determined by combinations of genotype and environment.

Some, but by no means all, of the critics of sociobiology deny this presupposition. Speaking for The Sociobiology Study Group of Science for the People, Lawrence Miller notes that:

> Once again, a dichotomy appears: genetics vs. environment. Once again, the dichotomy is a false one; both extremes present a *passive* view of humans as vehicles either for our genes or our environment. We view humans as *active* agents, striving to shape lives and destinies. Deterministic theories, whether genetic or environmental, serve to inculcate an ethos of passivity and thus render us susceptible to active manipulation by others.

Other critics of sociobiology make a similar point by emphasizing the role of human choice and intentionality (of freedom and responsibility) in the explanation of human behavior.[1] On the other side, Wilson, despite his advocacy of genetic determinism and his apparent materialist inclinations, sometimes speaks as if humans are capable of conscious choices that are "emergent" with respect to both genetic and environmental factors.[2]

The question of freedom vs. determinism, of the "emergent" quality of human choice, is extremely complicated and, in my view, it obscures the fundamental issue of the "nature-nurture" controversy: the extent to which human behavior is controlled by genetic programs rather than by culture and environment. Thus, for purposes of the present analysis, I shall presuppose the following *principle of determinism*: let G be the class of potential genotypes of some type of organism, let P be the class of its potential phenotypes, and let E be the class of potential environments in which the organism might develop; then their exists a function $f: GXE \rightarrow P$ such that for all $g \in G$ and $e \in E$, $f(g, e) = p$, where p is the phenotype that results when genotype g develops in environment e. The function f determines what is often referred to in population genetics as the *norm of reaction.*[3]

The fundamental epistemological problem posed by the sociobiological controversy is the following: given a particular phenotype p and the table of "reaction norms" $f(g, e)$, where g and e range over all possible genotypes and environments, to determine the degree to which p is caused or controlled by g versus the degree to which it is caused or controlled by e. Admittedly, we do not know (indeed, could not possibly know) the complete table of "reaction norms" for any phenotypic trait, but this is not the epistemologically significant point. The point is that even if we did have all that evidence, if we could systematically survey the complete table, there does not presently exist any adequate procedure for deciding the degree of genetic determination of any phenotypic trait. The problem is not simply that (as many critics of sociobiology have argued) there is at present no evidence that human social behavior is largely controlled by our genes; the problem is that disputants on neither side of that issue have a clear and consistent account of what would constitute such evidence.

There have been four distinguishable approaches in the sociobiological literature to defining a genetically controlled phenotypic trait: (1) the *capacity* account holds that genes have certain *capacities* or *potentialities* which set limits on the range of phenotypic traits that the organism can exhibit; (2) the *heritability* account holds that a phenotypic trait is genetically caused or controlled to the extent that the variation of that trait in a population is attributable to genetic (as distinct from environmental) variation in the population; (3) the *homeostasis* account holds that organisms are *homeostatic* mechanisms that are genetically (developmentally) *canalized* to produce specific phenotypic traits; and, (4) the *information* account holds that genotypes contain *information* in the form of programs that *code* for certain phenotypic traits. Although these are recognizably different interpretations of genetic determinism, they often appear in combination and various participants in the "nature-nurture" debate frequently equivocate among them.[4]

No one of these interpretations of genetic determinism is adequate, but I argue in the following that the third and fourth account, when combined in an appropriate way, provide the framework for a solution to the fundamental epistemological problem posed by sociology.[5] My analysis focuses upon the concept of a *program*. I contend that talk of "genetic codes," of "blueprints," and of "developmental programs", despite its strong intuitive appeal and heuristic value, is grounded in a metaphor that has not been given a sound epistemological foundation. My aim is to lay an adequate foundation for such talk. Many of the critics of sociobiology have emphasized the excessive reliance that advocates of genetic determinism have placed upon metaphorical ways of speaking.[6] In her excellent discussion of "Gene-Juggling," Mary

Midgley (1980) points out the dangers in such undisciplined use of metaphorical language, suggesting that, where metaphors are used in scientific discussions, they need to be "carefully pruned."[7]

Criticisms of sociobiologists' use of metaphorical language are well founded. but the problem is much more pervasive than the critics have realized. In particular, none of the participants in "the sociobiological debate" have exercised sufficient care in their use of the "information metaphor" when discussing the question of whether particular behaviors are "programmed by the genes." Genetic programs – especially "developmental programs" – are not directly accessible to observation: they must be inferred from empirical evidence regarding the regularities connecting genotypes, environments, and phenotypes. The fundamental problem is to determine the "genetic program" from the table of "reaction norms" (supposing that we had complete access to it). The solution to this problem requires a careful pruning of the information metaphor.

2. PROGRAMS AND THE INFORMATION METAPHOR

Recent contributions to the "nature-nurture" controversy illustrate the growing reliance upon the analogy to multipurpose computers, with their internally stored programs, as a device for conveying the nature of genetic determinism. The following passage is typical:

> Back to the wiring, and back to the new data we have about genetics and the control of living systems. Though we have known for many years how genetic systems worked, breaking the genetic code – the structure of the DNA molecule – allowed a rather hard-headed species to see with its own eyes how the life of an animal was controlled by facts of nature as comprehensible as a computer program. The genetic code *is* a program, and it *is* a way of transmitting a program from generation to generation.[8]

Among biologists, Ernst Mayr (1974a, 1974b) has given the most detailed and careful treatment of the concept of a "genetic program." Mayr has invoked the program analogy in connection with his attempt to legitimate biologists' use of "teleological" concepts and language. In order to avoid the connotations of conscious purpose and design in discussion of biological systems and processes, he has advocated (1974b, p. 102) that biologists substitute the terms 'teleonomy' and 'teleonomic' for the more traditional terms 'teleology' and 'teleological'. As Mayr recognizes, the proposed introduction of new language to convey the legitimate content of traditional modes of teleological descriptive can only be effective if these new terms have themselves been "purged" of implications of "conscious purpose or design."[9]

Like other writers who exploit the conceptual apparatus of computer programs, Mayr points to *information theory* as the parent discipline which gives those concepts their theoretical substance. However, there is a concealed ambiguity in the concept of information that renders that connection somewhat dubious. This ambiguity, which is simply a variant form of the ambiguity that undermines both the *capacity* and the *heritability* conceptions of "genetic determinism, can be made explicit by distinguishing between *descriptive information* and *prescriptive information*. Descriptive information is concerned exclusively with empirical probabilities – typically, probabilities understood as (actual or hypothetical) limiting relative frequencies.[10] For example, the "information content of a message" is a function of the difference between the prior and posterior probabilities of the outcome described by the message. If, in actual fact, 50% of all births are female, then a completely reliable message to the effect that a particular newborn is a female contains "one bit" of information. In a similar way, "channel capacity", "reduction of uncertainty", and other concepts central to information theory can be defined on the basis of the empirical probabilities of events in the actual world. The claim (Johnson, 1970, p. 1546) that, "The DNA in a mammalian cell has a capacity of about 2×10^{10} bits of information, which is equivalent to about 100 sets of the *Encyclopedia Britannica*", can be understood in this descriptive sense.

Prescriptive information, on the other hand, is a function not only of the actual empirical probabilities of events, but also of the "intended" probabilities of those events. This prescriptive sense of information is embodied in the metaphorical language of "blueprints" and "instructions". To put the difference another way, descriptive information is a function of what *is*, while prescriptive information is a function of what *ought to be* – the former is associated with descriptive regularities, the latter with prescriptive regularities. In his paper 'From Is to Ought', Anthony Flew (1967, p. 144) makes the general distinction in the following way:

> First, and certainly not confined to our present biological context, is the failure to distinguish two kinds of law of nature – or better, two senses of 'law of nature': the descriptive, in which such a law cannot have any genuine exceptions, since the occurrence of any event inconsistent with the truth of a proposed law constitutes a sufficient reason for failing the candidate; and the prescriptive, in which the occurrence of violations constitutes no reason at all for maintaining that the law originally propounded does not really obtain.

The regularities associated with blueprints and instructions are *prima facie* prescriptive in just the sense indicated by Flew: failure in practice to conform to the blueprint or to follow the instruction constitutes no reason at all for maintaining that the blueprint or instruction is not real. For nature, like the artisan to which Aristotle so often alludes, is capable of *mistakes*: there are transcription and translation "errors", messages that go awry, and blueprints that are misinterpreted.

As biologists have frequently noted, information in the descriptive sense is essentially a measure of complexity of structure: the more highly structured a system or process the more information it is capable of "storing or transmitting". However, the concept of structure (as applied to systems or processes) is not sufficient by itself to capture the teleological (or teleonomic) aspect of nature.[11] Mayr himself is certainly aware of the conceptual deficiency of a purely descriptive notion of information. Teleonomic processes are not simply highly structured processes – associated with a large "negative entropy" – they are in addition processes directed toward an end or goal. In his words (1974b, p. 98) "... *a teleonomic process or behavior is one which owes its goal-directedness to the operation of a program.*" The problem posed by the program metaphor, in other words, is that of translating the prescriptive connotations of information language into descriptive terminology. In particular, the problem can be stated as follows: given all of the descriptive laws governing a biological process or system – i.e., given the complete table of reaction norms – to characterize the program (or programs) governing the process or system. As Mayr argues, what is illegitimate in the use of teleologic language is the ascription to *nature* (or to natural processes or systems) of conscious design or intention. Such an ascription of conscious purpose is at the heart of Aristotle's comparison of nature to an artisan. It is this aspect of the information metaphor that requires pruning if the metaphor is to be an acceptable vehicle for an empiricist epistemology of sociobiology.

Stated in another way, the epistemological problem facing sociobiology – the problem which must be solved if we are to make empirical sense of reference to the degree of genetic determination (or control) of human social behavior – is that of providing an empirical procedure for *recognizing* biological programs. All that is empirically accessible, even in principle, is behavior – whether of atoms, molecules, organs, organisms, populations, or species. Thus purpose, wherever it exists, must be inferred on the basis of observable behavior under a whole range of possible circumstances (i.e., it must be inferred in some way or another from the table of reaction norms).

Mayr (1974a, 1974b) considers two general approaches to the recognition of programs: on one (*evolutionary*) interpretation of the term, a program is an historical entity whose content can only be understood on the basis of an understanding of how it developed; on the other (*functional*) interpretation, a program is an non-historical entity whose content is independent of the way it developed. Mayr argues strongly for the "functional" interpretation of the term 'program'.[12]

The contrast between the "evolutionary sense" and the "functional sense" of a program is crucial for purposes of the present discussion of genetic determinism. For, according to the "evolutionary" sense, the content of a program would be *essentially* connected with the history of natural selection by means of which the program had developed, and only *accidentally* connected with the actual functional regularities exhibited by the DNA; but the converse would hold for the "functional" sense of the term. In my judgment, Mayr's case for the functional interpretation of the term 'program' is overwhelming. The essential point is the following (see Mayr, 1974b, p. 103): if the term 'teleonomic' is understood in the evolutionary sense of program, then only those aspects of the teleonomic process that are the result of natural selection (that are *innate*) would fall under the definition. However, many aspects of teleonomic processes are *acquired* – they are learned during the lifetime of the organism. In Mayr's terminology, the evolutionary sense of program will only accommodate *closed programs*; it does not allow for the possibility of *open programs*. In effect, to adopt the evolutionary conception of teleonomy is to beg the question between the sociobiologists and their critics. Thus, the following analysis of biological programs, which is based upon Nagel's (1977) "system-property" account of goal-directedness, avoids the question begging features of the evolutionary account.

3. AN ANALYSIS OF THE CONCEPT OF A BIOLOGICAL PROGRAM

In his excellent article 'Teleology Revisited', Nagel (1977) considers three alternative explications of the notion of a goal-directed process: (1) the *intentional* view, according to which the ascription of goals to lower forms of life is to be understood as a metaphorical extension of the primary sense of goal-directedness, which is connected with conscious, intentional behavior; (2) the *program* view, according to which goal-directed processes are viewed as processes that are under control of a "program" (in Mayr's sense of the term); and, (3) the *system-property* view, according to which goal-directed processes are the result of organized systems that contain "homeostatic mechanisms" and "cybernetic feedback devices". Nagel rejects the first account of goal-directedness on the grounds that the analogy to intentional human behavior "contributes little to the clarification of the concept as it is used in biology." He rejects the second account on the grounds that the concept of a program is both too general and too vague to capture the sense of goal-directedness that motivated Mayr to introduce the term 'teleonomic process'. In criticizing the program view, Nagel argues (1977, p. 270) that "... if we do not have some idea of what goal-directed behavior is – an idea that, however vague it may be, is acquired and understood independently of Mayr's formal definition of teleonomic processes – the formal definition will neither convey the idea, nor provide instructions for applying it to biological phenomena, nor clarify the notion of being goal directed." Nagel argues that the *system-property* view, despite certain difficulties that it faces, provides a general framework within which to explicate the notion of goal-directedness. My strategy is to presuppose the *system-property* account as an analysis of goal directedness (or teleonomy) and in terms of it to explicate the notion of a "biological program", in effect turning Mayr's analysis on its head.

As Nagel (1977, p. 272) points out, the two characteristic features of goal-directed processes are *plasticity* – meaning that "the goal of such processes can generally be reached by the system following alternative paths or starting from different initial positions" – and *persistence* – meaning that

> the system is maintained in its goal-directed behavior as a result of changes occurring in the system that compensate for any disturbances taking place (provided these are not too great) either within or external to the system, disturbances which, were there no compensating changes elsewhere, would prevent the realization of the goal.

Systems which are organized so as to exhibit plasticity and persistence in their behavior are in a sense "programmed"; it is often possible to decompose such systems into a "program component" and a "programmed component", though it must be admitted that such a decomposition is neither absolute nor wholly objective, depending as it does upon the perspective and the pragmatic interests of the one who performs it.

Multipurpose computers, which are paradigm programmed devices, illustrate the pragmatic element in this decomposition. The typical microcomputer has a disk drive in which a magnetized "floppy disk" can be inserted. It is natural to view the floppy disk, with its particular pattern of magnetic impressions, as the program which controls the activity of the computer. However, the microcomputer will also have a "central processor" that constitutes a permanent part of its "program". This distinction is marked in "computer talk" by the terms 'software' and 'hardware', where the floppy disk would constitute "software" and the central processing "chip" would constitute "hardware". An important reason for identifying the program with "software" rather than "hardware" is that the former can easily be interchanged. In that respect computers are multipurpose: the goal toward which they are directed depends upon which "software" program is in control. The obvious biological analogue to a "software" program is DNA; the very notion of genetic determinism involves the question of whether different DNA (i.e., interchanging genotypes) would direct the "same organism" toward a different goal (phenotype).

Ease of interchangeability is one feature that distinguishes "software" from "hardware", but the terms are also used to convey a deeper and more problematic difference – a difference that is connected with a subtle ambiguity in the former term. On some occasions the term 'software' is used (in a *structural* sense) to refer to a physical structure – such as the magnetized floppy disk mentioned above – and on other occasions it is used (in a *functional-intentional* sense) to refer to a particular function or end which the physical structure (the magnetized floppy disk) was intended to serve. When we speak of programs written in particular programming languages (such as FORTRAN, BASIC, etc.) the reference is to the *functional-intentional* sense of 'software'. When 'software' is used in this sense the floppy disk, as a physical structure, becomes "hardware". The *functional-intentional* sense of 'software' is problematic precisely because it invokes the *prescriptive* sense of information. In fact, this sense of 'software' acquires its meaning from the same metaphorical extension of human *intentionality* that Nagel rejected on the grounds that it served little to clarify the concept of goal-directedness as it is used in biology. So far as its contribution to the resolution of the "nature-nurture" controversy is concerned, the functional-intentional sense of 'software' is an *epistemological cul-de-sac.*[13]

Thus, in the present analysis we will use the term 'program' in a strictly *structural* sense: a software program will consist of some "easily interchanged" component of a system that plays a special role in the plasticity and persistence of that system in maintaining the goal toward which it is directed. Again, the obvious biological paradigm is DNA, viewed as a physical-chemical structure. The decision to use the term 'program' in a strictly structural sense has two important implications. In the first place, the concept of a program is relative; specifically, a program is the program that it is relative to the system of which it is a component. For instance, a particular floppy disk inserted in an IBM PC may constitute a program for that device, while the very same disk inserted in a different computer may not constitute a program at all (because it does not enable the computer to carry out a goal-directed process) or it *may constitute an entirely different program* (because as a component of the second computer it contributes to a very different goal).

The point is that programs have no identity as programs apart from the systems of which they are components. The "fact" that the programmer *intended* the floppy disk, with its particular arrangement of magnetic tracks, to convert the multipurpose computer into a word processor is irrelevant: if it happened (by a "miracle") that that very disk converted a second computer into a "financial planner", then relative to that computer the disk would constitute a "financial planning program" rather than a "word processing program." As the following, admittedly fanciful, example illustrates, the *relativity* of programs is crucial to the correct analysis of "biological programs". Consider a mature chimpanzee in which all of the DNA (in both germ cells and somatic cells) has been replaced by human DNA. Presumably, such an organism would not continue to "function", but if it did, and if it reproduced with its "own kind" (i.e., another mature "chimpanzee" with human DNA), the result might well be an animal that differed in significant respects from both chimpanzees and humans. Despite the "universality of the genetic code", it is a fact (indeed a "conceptual fact" more than an "empirical fact") that the "meaning" of DNA depends upon the context in which that DNA is embedded (the system of which it is a component). At least in principle (and, as I will argue below, in actual fact) the "same" genetic program (genotype) can "mean" different things in different biological systems.

There is a second important implication of the decision to use the term program in a strictly *structural* sense: in a fundamental respect, programs are always "programs for stability" rather than "programs for change." A program is a structural component of a goal-directed system – a component that plays a special role in the plastic and persistent process whereby the system maintains its goal state. A program becomes such only when it can be viewed as that sort of stabilizing component of a goal-directed system. Of course, goal directed systems undergo change and development – change and development which is under program control – but the claim that a particular process (or development) is under program control presupposes the possible existence of a stable state, the maintenance of which is the goal of the system.[14]

The process of reproduction is the most characteristic feature of living systems and, in light of the point made in the preceding paragraph, it is the feature which licenses talk of "biological programs". For, reproduction may be viewed as a type of stability: a system which reproduces itself is, in effect, a system which maintains itself. Of course, this is true only to the extent that the "offspring" are *the same* as the "parent". Talk of *reproduction* illustrates another important point about the concept of a program: there is necessarily a process of abstraction involved in the specification of the goal of a goal-directed system. Stability and change are relative notions; an organism can maintain a stable structure, for example, despite the fact that there is a complete "turnover" in the molecules that constitute its body. The important point to recognize here is that we can think of reproductive systems as being under program control only to the extent to which we view the offspring as constituting *the same system* as the parent; for only in that case can we view the reproductive process as the maintenance of a system in a stable goal state. This requirement poses a peculiar difficulty for the analysis of sexually reproducing systems. For that reason, it is useful to first apply the concept of a program to "reproductive systems" which do not involve the "complications of sexuality."

A *biological program g, relative to a system S*, is a physical component of *S* which plays a special role in enabling *S* to reproduce (or maintain) itself under a range *E* of environmental conditions. The "special role" of *g* is to be understood as involving the plasticity and persistence of the process whereby reproduction takes place under a variety of environmental conditions. It is important to note in connection with this definition that reproduction of the system *S* involves reproduction of the program *g*, since that program is, itself, a structural component of the system. In this way, reproduction can be seen as maintenance, for the reproduced system is in exactly the state required to again reproduce itself, provided that it continues to "dwell" in the proper environment (one of the environments in the range *E*). The computer analogy may help to draw out the implications of this conception of a "biological program." Imagine a computer which will, upon insertion of a particular floppy disk, proceed to reproduce itself: more specifically, a computer which will, by following different paths under different environmental conditions (involving the "external inputs" to the hardware device), proceed to manufacture an "identical" computer in which exactly the same program (floppy disk) is a component. Such a computer would constitute a system under the control of a "biological program".

Again, it should be noted in this example that the floppy disk is a program *relative to* the hardware device in which it is a component. The "function" of the program is to manufacture a particular computer, *but only by virtue of the fact that it is a component of that computer in the first place*: as a component of a different computer it might have the function of manufacturing that different computer. Applied to asexual organisms, this definition entails that biological programs control a cloning process, wherein the progeny are genetically and phenotypically identical to the parent organism. According to this account, evolution can only occur by virtue of imperfections in biological programs, for only systems that fail (in the reproductive process) to maintain themselves in the goal state provide "raw material" upon which natural selection can operate.[15]

The definition of "biological program" proposed above seems particularly suited to asexual organisms; it seems less well suited to organisms which reproduce sexually, since such organisms do not transmit "identical" copies of their genotype (because of recombination) and do not reproduce "identical" phenotypes either. As already mentioned, there is necessarily a process of abstraction involved in the specification of the goal of a goal-directed system (or in the determination of whether two temporal stages are stages of the "same system") and talk of "reproduction" always involves an implicit *similarity relation*, since the offspring are always *similar to* rather than *identical with* their parents. So one might extend the definition of biological program to sexual organisms by identifying the genotype (program) with that part of the DNA that is "invariant" across members of the breeding population, and by identifying the phenotype (goal-directed system) with the corresponding invariant features of the organisms' physiology and behavior. However, this extension of the definition would still not account for the distinctive role of sexuality in biological reproduction.

An intuitively more plausible way of extending the definition to sexually reproducing organisms is to think of the biological program as a characteristic of the breeding population (or species) rather than as a characteristic of the individual organisms themselves. On this interpretation, the biological program would be identified with the *statistical distribution of genotypes* in the population, and the goal-directed system would be identified with the *statistical distribution of phenotypes* in the population. A case for the existence of a biological program could then be mounted by producing empirical evidence to show that a specific population "would manage" in a plastic and persistent manner to preserve those statistical distributions of genotypes and phenotypes in the face of changes in the larger ecosystem (within some range E). On this interpretation reproduction is essentially a continuous process and the distinction between "reproduction" and "maintenance" vanishes.[16]

For purposes of this paper, whose ultimate concern is the question of genetic determinism as it arises in the epistemology of sociobiology, the most important feature of Mayr's definition of teleonomy is his distinction between *open* and *closed* programs. This distinction was intended to do the work of the vague distinction between *innate* and *acquired* traits; for, according to Mayr (1974a, p. 695), what we mean by saying that a trait is innate (acquired) is that it is under control of a closed (open) program. As he explains the difference (1974b, p. 103),

> . . . programs which control teleonomic processes in organisms are either entirely laid down in the DNA of the genotype ('closed programs') or are constituted in such a way that they can incorporate additional information ('open programs') . . . acquired through learning, conditioning or through other experience.

In his critique of Mayr's analysis of teleonomy, Nagel (1977, p. 270) argues that ". . . it is at present difficult if not impossible to ascertain by actually examining a *program*, as distinct from the overt *process* it controls, whether the program is closed or open." I believe that Nagel's criticism of Mayr's way of drawing this important distinction can be strengthened in two ways: (1) since programs are, by definition, identifiable only as components of particular systems, it is impossible *in principle* (and not merely *in practice*) to determine what a program *is* (let alone whether it is *open* or *closed*) by examining the program in isolation from the system it controls; and (2) it is not evident from Mayr's discussion of the distinction how one would decide, even if one had access to the entire table of reaction norms, whether a program (viewed as a component of a particular system) is open or closed.[17]

Mayr's way of drawing the distinction between closed and open programs has a strong intuitive appeal: closed programs are entirely self-contained; their "meaning" is independent of the context in which they are embedded. However, that intuition is incompatible with the "relativity of programs." Pure DNA extracted from a few human genomes and sprinkled upon the ground will not repeople an uninhabited planet.[18] Human DNA constitutes a biological program only when embedded in an appropriate system (i.e., when embedded in an appropriate distribution of phenotypes), but it may constitute different programs when embedded in different systems, and in that respect the program cannot be "entirely laid down in the DNA".[19] Nevertheless, Mayr's terms 'open' and 'closed' point to an important distinction; the question is how to draw it. The biological program for the characteristic human physique is evidently "closed" in a way that the biological program for speaking the English language is not closed. Notice, however, that even in the latter case it is proper to speak of a "genetic program": a given distribution of genotypes, embedded in a system of phenotypes whose behavioral characteristics include English speech, will, in a plastic and persistent manner, maintain that system of "English speaking phenotypes" (with the "same" distribution of genotypes) under a variety of environmental conditions. In that sense human DNA constitutes a biological program for the English language. Of course, that same distribution of genotypes, embedded in a system of "German speaking phenotypes" (or a system of "Swahili speaking phenotypes"), may also constitute a biological program for the German language (or for the Swahili language), because the system, under control of that distribution of DNA, maintains (reproduces) the goal state of German (Swahili) speech.

This example of human language behavior indicates the intuitive basis of the distinction between open and closed programs. The DNA program for the English language is "open" in the sense that the biological system (the system relative to which that program is the program that it is) is not "highly restricted in scope." As noted above, the distinction between program and system is somewhat arbitrary (it has a pragmatic aspect); for similar reasons, the distinction between system and environment also has a pragmatic aspect. If we construe the term 'phenotype' in a narrow sense, so that the phenotype involves *physiological* but not *behavioral* traits, then the system consisting of a particular statistical distribution of phenotypes (in that narrow sense) will not be programmed for "the English language" (or for any other specific language). The "goal state" of this "physiological system" (in which 'phenotype' is narrowly construed) is broader than the goal state of the "behavioral system" (in which "phenotype' is broadly construed). In effect, factors (such as speech behavior) which are part of the "behavioral system" are included in the environment of the "physiological system". Thus, the DNA program for the "physiological system" is more nearly closed than the DNA program for the "behavioral system".

Let us say that a goal-directed system S is *contained in* a goal-directed system S' provided that: (i) every "goal state" of S is a "goal state of S' and (ii) the range of environments will respect to which S is goal directed is included in the range of environments with respect to which S' is goal directed. In this sense the "behavioral system" is contained in the "physiological system": for instance, (i) all those cases of English speaking cultures are cases of (physiologically) human populations and (ii) all of those environments in which cultures can maintain the English language are environments in which human populations can maintain the human physiology. Now, we can say that a program g is *more closed* as a program for system S' than it is as a program for system S' if and only if S is contained in S'. The scope of this definition is rather limited; in particular, it provides no direct way to determine whether one program g is more or less closed than another program g' (no direct way to determined, for instance, whether the genetic program for mating behavior in wild chimpanzees is more or less closed than the genetic program for mating behavior in existing human populations). However, such questions can be answered in a rough sort of way by looking for correspondences or analogies between the features of distinct biological systems.

For example, it might be that a human genetic program g (a particular statistical distribution of human genotypes) embedded in a biological system S whose goal state was determined by a characteristic distribution of human physiological phenotypes, would *contain* each of a variety of biological systems $S_1, S_2, \ldots, S_k$ (representing different cultures), where the goal state of each S_i was determined by a distinctive pattern of mating behavior (in addition to the common physiological phenotype), and where, furthermore, g was a program for each system S_i. In this event, we would say that g was a program for a variety of systems of human mating behavior, and the more distinctive those systems were (the more "specialized" the behavior exhibited by the systems) the more *open* the biological program for human mating behavior. If it turned out that there were fewer and "less specialized" (goal-directed) systems of mating behavior for chimpanzees, then we could say (by virtue of the analogy between the physiological and behavioral systems that are characteristic of humans and chimpanzees, respectively) that the biological program for mating in chimpanzees was more closed than that for mating in humans.

To see how the present distinction between open and closed programs can be applied to traditional problems of ethology, consider the following questions which Mayr (1974a, p. 696) raises in his discussion of 'Behavior Programs and Evolutionary Strategies':

> ... Do all animals have *innate releasing mechanisms* that respond to specific sign stimuli (releasers), resulting in appropriate responses (displays, etc.)? Is a given releasing mechanism controlled by the genetic program of the species (innate), or is it individually acquired either by gradual learning or by a specially rapid learning process (imprinting)? Is the releasing mechanism located in the central nervious system or is it part of the screening mechanisms in the peripheral system (including the sense organs)?

From the present perspective, the question of whether "a given releasing mechanism is controlled by the genetic program of the species ..." is somewhat misleading. Presumably, there will always be some biological system S which is "controlled by the genetic program of the species" and which "invariably" exhibits the responses (displays, etc.) associated with the *releasing mechanism*. For instance, it is obvious that "appropriate mating behavior" must be an "invariable" characteristic of any biological system that successfully maintains (reproduces) itself. If specific displays of the male are releasing mechanisms for the female, then that display behavior must occur regularly as an aspect of the goal-directed system. In that respect, it will always be true that the releasing mechanisms is controlled by the genetic program of the species.

Given that there will always exist a biological system S (a particular distribution of phenotypes) controlled by the genetic program of the species and exhibiting the appropriate releasing mechanism, the question of whether that releasing mechanism is *innate* or *acquired* can perhaps be understood as follows: Does there exist a biological system S' which *contains* S, which is also controlled by the genetic program of the species, and in which the goal state can be specified without reference to the given releasing mechanisms? In the event that there is such a "containing" biological system the releasing mechanism is *innate*, otherwise it is *acquired*.[20] The difference is that some traits must be included in the specification of the goal state of the system in order that the system should be goal directed (let us call these the *essential traits*), while other traits (the *accidental traits*) are present in the system by virtue of their "teleomatic" (lawlike) associations with the essential traits. Take language behavior as an example. We can specify a biological system (a distribution of phenotypes) which is goal directed for English language behavior; however, language behavior is, itself, an essential aspect of the goal state of such a system. That is to say, it is only systems that already display English language behavior which will, upon insertion of a given genetic program (distribution of human genotypes), exhibit English language behavior. In that sense English language behavior is acquired. On the other hand, skin color is "genetic" because it is possible to specify a biological system (where skin color is not one of its essential traits) for which a particular distribution of "racially homogeneous" human genomes is a program and which will "upon insertion of that program" maintain a homogeneous skin color. In effect, the language behavior of the parents (the language behavior of an earlier stage of the biological system) is essential to the language behavior of the offspring (the language behavior of a later stage of the biological system), while the skin color of the parents (skin color of the earlier stage) is irrelevant to the skin color of the offspring (skin color of the later stage).[21]

In regard to the implications of the preceding discussion for "the socio-biological debate", I must issue a caveat. Following the standard practice of ethologists, I have distinguished between physiological phenotypes and behavioral phenotypes, or between physiological and behavioral traits of a goal-directed biological system. However, there are strong reasons to suppose that behavioral traits are simply *teleomatic* consequences of (physiological) brain states. It follows, for example, that English language *behavior* is not an *essential* aspect of the biological system (distribution of phenotypes) that is programmed (by any characteristic distribution of human genotypes) for English language behavior. All that is required is that the system exhibit some appropriate distribution of brain states and the appropriate language behavior will follow. So rather than distinguishing between physiological (or structural) phenotypes and behavioral phenotypes we need to distinguish between parts of the physiology of the brain, or between different levels of *Programs of the Brain* (see Young, 1978). The claim that a trait is "innate" can then be understood as a claim that it is a teleomatic consequence of low level brain processes.[22]

4. CONCLUSION

My aim in this paper has been to sketch an analysis of the concept of a biological program and in terms of that analysis to contribute to the resolution or dissolution of the dispute over genetic determinism, a dispute which has been at the center of "the sociobiological debate". My analysis has been guided by traditional empiricist epistemological principles in the following respect: an adequate explication of 'genetic determinism' should enable one to answer questions about the "degree of genetic causation or control" on the basis of evidence concerning ("lawful") empirical regularities in the actual world. Roughly speaking, the relevant regularities are those connecting genotypes, phenotypes, and environment and are given by the *reaction norms*, which map specific combinations of genotype g and environment e into a specific phenotype p. The "actual world" is governed by *descriptive* laws which specify what *is* or *might be* the case as distinct from *prescriptive* laws which specify what *ought to be* the case.[23]

I have argued that the familiar "information metaphor", involving talk of codes, blueprints, and instructions, has led (or enabled) biologists and others involved in the "nature-nurture" controversy to invoke *prescriptive* "regularities" in their account of the biological world. It is this prescriptive aspect of teleological language – language which derives its meaning from a metaphorical extension of the notion of *intentional* human action – that led Mayr to introduce the term 'teleonomy'. Recognizing that teleonomic processes involved something more than "mere empirical regularity" – something more than *teleomatic processes* – Mayr invoked the concept of a program. However, the term 'program' must, itself, be rid of prescriptive implications. Following Nagel (1977), I have attempted to characterize biological programs in terms of the notion of a goal-directed system. Roughly speaking, a biological program is a "structural" component of a goal directed system which "controls" the plastic and persistent process whereby the system maintains its goal state in a range of external environments.

There are two important consequences of this definition that have not heretofore been sufficiently appreciated yet which are, in my judgment, of great significance for the "nature-nurture" controversy and the question of genetic determinism. In the first place, a biological program is a *relation* between two physical structures: the program structure and a goal-directed system which is programmed by that program structure. Thus, talk of DNA as a program is elliptic; the same genotype embedded in distinct phenotypes may constitute distinct programs. As a corollary, the question "Is a particular trait under genetic control?" or the question "To what extent is a particular trait under genetic control?" has no obvious meaning. The intuitions behind talk of "degrees of genetic control" can be explicated in terms of the *containment* relation between biological systems. Systems whose goal states are more narrow ("specialized") are contained in those whose goal states are broader ("less specialized"), and questions about the degree of genetic control of traits are really questions about the degree of generality of the (essential) goal states of biological systems.

The greater the range of specific states (values of essential phenotypic traits) that are within the general goal state of the system, the greater the degree of control of the genetic program. For example, English language behavior does not exhibit a high degree of genetic control (by contrast, say, with "skin color") because language behavior (or the corresponding "brain program") constitutes an essential aspect of the goal state of a biological system which is genetically programmed to maintain such behavior. In other words, it is only if DNA is embedded in a population of organisms who "already speak English" that that DNA will program the population to exhibit English language behavior. Note that a trait would be "completely genetic" only if there were no essential constraints upon the goal state of the containing system. For instance, eye color would be completely genetic only if the genotype led in a completely plastic and persistent manner to eyes of a given color, so that even a genotype "sprinkled upon the ground" would produce eyes of that color. Of course, in that sense there are no completely genetic traits.

The second consequence of the present analysis of the term 'program' is that programs are necessarily identified by the contribution which they make to the mainenance of stability; there is no such thing as a "biological program for change". Evolution, operating through the mechanism of natural selection, rests upon the defects of biological programs – upon the failure of systems under the control of DNA programs to maintain biological systems within their goal states. The evidence for the existence of a biological program is the plasticity and persistence of the process whereby the biological system which it controls maintains its goal state under a wide range of environmental conditions. The evidence for the failure of a biological program is the breakdown of those "homeostatic mechanisms" and "feedback devices", breakdowns which typically occur at the outer limits of the environments in which the program effectively maintains the biological system.[24]

The efforts of sociobiologists – and Wilson (1978), in particular – to draw inferences about *human nature* from the behavior of past and of present "hunter-gatherer cultures" clearly go against the grain of the present analysis. There is little doubt that the present human genes (statistical distribution of genotypes) constitute a program for hunter-gatherer culture. It constitutes such a program because if those genes *were to be* interchanged with the genes of a population whose behavior was characteristic of such a culture, that population would persist in its "hunter-gatherer" behavior. In that respect, it is the *nature* of humans to be "hunter-gatherers". However, it is equally true (we trust) that the present human genes constitute a program for the culture of 20th-century industrial democracy, in the sense that those genes embedded in a population which practices such a culture will, under a wide range of environmental conditions, contribute to the maintenance of that culture. In that respect, it is the *nature* of humans to be "civilized".

Perhaps then the debate over human nature comes down to an empirical question of the relative degree of perfection of the *different programs* constituted by human DNA. But in that case the sociobiologists clearly have the losing end of the argument; indeed, it is quite clear that within the range of environments that have actually existed on earth, the "hunter-gatherer" program is defective. For, otherwise, we would still be participating in a "hunter-gatherer" culture and the whole "sociobiological debate" would never have arisen. The question of whether the biological program that is associated with our present culture is defective is more difficult; if the pessimists are right it is defective, if the optimists are right it may succeed. It is interesting that much of the modern *Utopian* literature (for example, Skinner's *Walden II*) can be understood as claiming that the present human genes constitute a biological program relative to some nonexistent biological system. It is tempting, on occasion, to interpret the remarks of the sociobiologists as those of "reactionary Utopians", with the "hunter-gatherer" culture as the nonexistent biological system that represents Utiopia. In any case, what Utopias (from Plato's *Republic* to the present) have often preached is that "if once we could realize the Utopian system (culture), human nature would enable us to maintain it."

According to the present analysis, therefore, the question of whether human genes constitute a program for a given biological system is strictly empirical (or descriptive); its answer depends upon the plasticity and persistence of the programmed process whereby that system maintains itself in its goal state in response to changes in its external environment. On the other hand, the question of which of the many biological systems for which the human genes constitute a program (or near program) really reflects human nature is not strictly empirical. It is a prescriptive (value) question to which biology – including the science of sociobiology – can provide no objective answer.

Michigan State University

NOTES

[1] See, for example, Barnett (1980, p. 151).

[2] See his (1978, Ch. 4). Actually, Wilson's position on the question of human freedom and responsibility is ambivalent and perhaps incoherent. On the one hand, he equates free choice with the inability to predict (we are free if our behavior cannot be predicted by ourselves or others), which makes it a relative notion; on the other hand he suggests (1978, p. 100) that, through the self knowledge acquired from the science of sociobiology, "We can hope to decide more judiciously which of the elements of human nature to cultivate and which to subvert, which to take open pleasure with and which to handle with care."

[3] See Lewontin (1974, p. 404). The present *principle of determinism* is not intended to foreclose the question of indeterminism: the question of whether biological processes are irreducibly stochastic. In that event the *norm of reaction* would yield a statistical table, such that, for $g \in G$ and $e \in E$, $f(g, e)$ is a probability distribution over the class P of possible phenotypes. Let us say that $f^*(g, e, p)$ is the propensity for genotype g and environment e to produce phenotype p. For the present purposes, these propensities can be understood in a modal sense as proposed by Fetzer (1981) and others. To allow

for the possibility that for particular values of *g*, *e* and *p*, $0 \neq f^*(g, e, p) \neq 1$ is, in effect, to recognize that the universe may be irreducibly stochastic; it is not to make any assumptions one way or the other regarding the question of free will. For purposes of the present analysis, nothing much turns on whether the phenotypic propensities are universal or statistical. The point is that the propensity $f^*(g, e, p)$ does not depend upon anything beyond the genotype *g* and the exhaustively described enviornment *e*.

[4] Wilson (1978), for example, invokes each of the first three approaches at one point or another in his treatise *On Human Nature*.

[5] Due to limitations of space, sections of the present paper dealing with the *capacity* approach and the *heritability* approach to defining "degree of genetic determinism" had to be omitted. The principal thesis of those omitted sections is that these approaches, like the "information-theoretic" approach which is the focus of the remaining sections, are guilty of confusing *prescriptive* and *descriptive* questions. As a result they amount to an attempt to derive *ought* from *is*.

[6] See, for example, the Sociobiology Study Group of Science for the People (1976), p. 263), Miller (1976, p. 276), and Burian (1978, p. 378).

[7] Midgley (1980, p. 118) notes that: "To understand how metaphors can properly be used in scientific writing, we must get straight a fundamental point about the relation between metaphors and models. Every metaphor suggests a model; indeed, a model is itself a metaphor, *but one which has been carefully pruned*. Certain branches of it are safe; others are not, and it is the first business of somebody who proposes a new model to make this distinction clear. Once this is done, the unusable parts of the original metaphor must be sharply avoided; it is no longer legitimate to use them simply as stylistic devices."

Midgley uses the metaphor of *mechanism* to illustrate her point, noting that it has been successfully stripped of its implication of an inventor or maker. Unfortunately, the metaphor of *information* has not yet been successfully stripped of its implication of conscious purpose or intentionality.

[8] Tiger and Fox (1971, p. 58).

[9] Thus, Mayr notes (1974b, p. 102) that: "The key word in my definition of 'teleonomic' is the term 'program'. Someone might claim that the difficulties of an acceptable definition for teleological language in biology had simply been transferred to the term 'program'. This is not a legitimate object, because it neglects to recognize that, regardless of its particular definition, a program is (1) something material, and (2) it exists prior to the initiation of the teleonomic process. Hence, it is consistent with a causal explanation.

Nevertheless, it must be admitted that the concept 'program' is so new that the diversity of meanings of this term has not yet been fully explored. The term is taken from the language of information theory. A computer may act purposefully when given appropriate programmed instructions. Tentatively program might be defined as *coded or prearranged information that controls a process (or behavior) leading it toward a given end*."

[10] However, nothing of significance turns upon the particular interpretation of probability that is chosen. In the present sense of the term 'descriptive', even those conceptions of probability – such as the "propensity" account advocated by Fetzer (1981) – which rely upon modal "possible worlds" talk are purely *descriptive* as opposed to *prescriptive*.

[11] This deficiency in traditional descriptive information theory is noted in Johnson's

critique of 'Information Theory in Biology after 18 Years.' He argues (1970, p. 1550) that: "Information theory has failed thus far to provide a useful calculus for experimental biology. This failure can be laid to the fact that information theory, in its present form, quantifies negative entropy, an extensive property. In thermodynamics the extensive property, entropy, must be modified by an intensive property, temperature, in order to signify the feasibility of a process. Similarly, in biology the extensive property, information or negative entropy, can have significance for the feasibility of a living process only if it is modulated by an intensive property which indicates the biological relevance or purposefulness of that information."

[12] For example, Mayr notes (1974b, p. 99) that: "My definition of 'teleonomic' has been labeled by Hull (1973) as a 'historic definition'. Such a designation is rather misleading. Although the genetic program (as well as its individually acquired components) originated in the past, this history is completely irrelevant for the functional analysis of a given teleonomic process. For this it is entirely sufficient to know that a 'program' exists which is causally responsible for the teleonomic nature of a goal-directed process. Whether this program had originated through a lucky macromutation (as Richard Goldschmidt had conceived possible) or through a slow process of gradual selection, or even through individual learning or conditioning in open programs, is quite immaterial for the class of a process as 'teleonomic.'

[13] I would have made the contrast between *structural* and *functional* (rather than *functional-intentional*) programs, except that the term 'functional' (like the terms 'natural' and 'normal') is ambiguous. In its *descriptive* sense, a function is simply a structural regularity (mathematics deals with functions in this sense); in its *prescriptive* sense a function involves intentionality or conscious purpose. Mayr clearly has the descriptive sense of function in mind in contrasting the *evolutionary-historical* and the *functional* sense of programs. It is interesting that the *evolutionary-historical* sense of function is quite close in some respects to the *functional-intentional* sense: if we were to attribute conscious purpose to the evolutionary process, the former sense would become a species of the latter.

[14] This is yet another respect in which (structural) programs are unlike intentions: we can intend to change something or to cause it to develop without presupposing some fixed end toward which that change or development is directed. The notion that programs are always "programs for stability" also strengths Mayr's argument (1974b, pp. 95–97) against a "teleological determination of evolution."

[15] Again, it is important to note that it does not make sense to suppose that biological systems have been programmed to fail, so in that respect there cannot be a "program for evolution." It is also worth noting in passing that this account allows for the occurrence of rapid evolutionary change in situations in which radical changes in the environment lead to the replacement of one program by another – as suggested by the *punctuated-equilibrium* model of Eldredge and Gould (1972). In the stressed environment one might expect that latent genetic variation (differences in the genetic programs of organisms that did not produce phenotypic differences) would lead to a proliferation of phenotypes (representing the variety of ways in which the program could fail); these distinct phenotypes would then provide extensive raw material upon which natural selection could act. Although most of the programs (DNA structures) which led to a constant phenotype, say p, under "normal" environmental conditions E would not be programs at all under the radically changed conditions E', some few of them might

retain their status as programs – but as programs for a radically different phenotype p' (programs for p' because, when embedded in E', they reproduced p' in a plastic and persistent manner under the range of environmental conditions E'). In such a situation the DNA program would have undergone a "radical change in meaning" in a single generation and one would have an instance of "punctuated equilibrium." Also, compare Hull (1980, p. 81).

[16] One advantage of this proposal is that it adds plausibility to the claim of David Hull (1978, 1980) and others that species (breeding populations?) are, themselves, individuals (rather than just classes of organisms). They are individuals because they are goal-directed systems regulated by biological programs.

[17] Mayr (1974a, p. 700) discusses several kinds of evidence that might "help to decide to what extent (a) genetic program is open or closed." However, the factors cited, though they are intuitively plausible as indicators, do not provide any sort of definition of the "relative degree of openness of distinct genetic programs." I would see the sort of intuitive factors mentioned by Mayr as one test of the plausibility of the present analysis.

[18] See James King (1980, p. 97).

[19] This point about the relativity of the information contained in a program, and the consequence of this relativity for the innate/acquired distinction, is related to a point that William Wimsatt has recently emphasized in a discussion of the concept of "innateness." In Wimsatt's words (1984):

> Given that fitness is a relation between organism and environment, then the information relevant to fitness is contained in the relation between organism and environment, and sufficiently stable features of either or of their relationship can play a heritable and generative role. It then seems plausible to say that what is innate is their interaction, and that "genetic information", rather than being a property of the DNA, is functional heritable information deriving from the interaction of both genes and environment.

The present analysis of information differs from Wimsatt's only in emphasizing that information is associated with the persistence and plasticity of a goal-directed system. Thus, there is information in the interaction between organism and environment because there are mechanisms which maintain the stability of that interaction.

[20] Actually, this distinction must be understood in a comparative sense: phenotypic traits are *more* or *less* innate.

[21] Skin color of the parents is irrelevant in the sense that if the DNA of a population of phenotypes which is homogeneous for one skin color were systematically interchanged with "DNA which is homogenous for a different skin color," the latter skin color would be exhibited in the offspring, and would become a "fixed point" of the biological system. There is no comparable way of making sense of the expression "DNA which is homogeneous for English language behavior."

[22] See, for example, Wilson's discussion (1978, pp. 42–43) of the role of the frontal lobes of the brain versus that of the limbic system.

[23] Note that I am here using the term 'actual world' in a sense that includes *possible worlds*. Reaction norms are to be understood in a modal (subjunctive) sense: phenotype *p would result* if genotype *g were to be* embedded in environment *e*. The relevant contrast is not between *actual* worlds and *possible* worlds but rather between *actual worlds* and *ideal worlds*, where ideal worlds are special kinds of possible worlds.

[24] See Hull's discussion (1980, p. 81) of Mayr's account of "genetic diversity and the

unity of the genotype." One can think of the genotype as having a unity because the "statistical distribution of genotypes" constitutes a program for the system comprised by the "statistical distribution of phenotypes."

REFERENCES

Barnett, S. A.: 1980, 'Biological Determinism and the Tasmanian Native Hen'. (In Montagu (1980), pp. 135–157.)

Burian, R. M.: 1978, 'A Methodological Critique of Sociobiology'. (In Caplan (1978), pp. 376–395.)

Caplan, A. L. (ed.): 1978, *The Sociobiology Debate*, Harper and Row, New York.

Eldredge, N. and Gould, S. J.: 1972, 'Punctuated Equilibria: An Alternative to Phyletic Gradualism', in T. J. M. Schopf (ed.), *Models in Paleobiology*, Freeman, Cooper and Company, San Francisco, pp. 82–115.

Fetzer, J. H.: 1981, *Scientific Knowledge*, D. Reidel, Dordrecht.

Flew, A.: 1967, 'Evolutionary Ethics', in W. Hudson (ed.), *New Studies in Ethics*, Macmillan and Co., London, pp. 31–51. (As reprinted in Caplan, 1978, pp. 142–162.)

Hull, D.: 1978, 'A Matter of Individuality', *Philosophy of Science* **45**, 335–360.

Hull, D.: 1980, 'The Herd as Means', in *PSA 1980*, Vol. 2, ed. P. D. Asquith and R. N. Giere (eds.), Philosophy of Science Association, East Lansing, MI, pp. 73–92.

Johnson, H. A.: 1970, 'Information Theory in Biology after 18 Years', *Science* **168**, 1545–1550.

King, J. C.: 1980, 'The Genetics of Sociobiology'. (In Montagu, 1980, pp. 82–107.)

Lewontin, R. C.: 1974, 'The Analysis of Variance and the Analysis of Causes', *American Journal of Human Genetics* **26**, 400–411.

Mayr, E.: 1974a, 'Behavior Programs and Evolutionary Strategies', *American Scientist* **62**, 650–659. In: *Evolution and the Diversity of Life: Selected Essays*, Cambridge University Press, Cambridge, pp. 694–711.

Mayr, E.: 1974b, 'Teleological and Teleonomic, A New Analysis', in R. S. Cohen and M. W. Wartofsky (eds.), *Methodological and Historical Essays in the Natural and Social Sciences*. (Boston Studies in Philosophy of Science Vol. XIV), D. Reidel, Dordrecht, pp. 91–117.

Midgley, M.: 1980, 'Gene-Juggling', *Philosophy* **54**. (As reprinted in Montagu, 1980, pp. 108–134.)

Miller, G. M.: 1976, 'Philosophy, Dichotomies, and Sociobiology', *Hastings Center Report* (Oct. 24), pp. 20–25. (As reprinted in Caplan, 1978, pp. 319–324.)

Montagu, A.: 1980, *Sociobiology Examined*, Oxford University Press, Oxford.

Nagel, E.: 1977, 'Teleology Revisited', *Journal of Philosophy* **74**, 261–301.

Sociobiology Study Group of Science for the People: 1976, 'Sociobiology – Another Biological Determinism', *Bioscience* **26**. (As Reprinted in Caplan, 1978, pp. 280–290.

Tiger, L. and Fox, R.: 1971, *The Imperial Animal*, Holt, Rinehart and Winston, New York. (As reprinted in Caplan, 1978, pp. 57–63.)

Young, J. Z.: 1978, *Programs of the Brain*, Oxford University Press, Oxford.

Wilson, E. O.: 1978, *On Human Nature*, Harvard University Press, Cambridge.

Wimsett, W.: 1984, 'Von Baer's Law of Development, Generative Entrenchment, and Scientific Change', *Philosophy of Science* **51**.

ROBERT N. BRANDON

PHENOTYPIC PLASTICITY, CULTURAL TRANSMISSION, AND HUMAN SOCIOBIOLOGY

ABSTRACT. Two major criticisms of human sociobiology are: (1) Biological determinism, the view that behavior is determined by genes, is wrong when applied to humans; and (2) Culture makes man unique in the animal world, so one cannot extrapolate from the sociobiology of non-cultured animals to man. I argue that the two criticisms are related in that a range of the conditions that favor increased phenotypic plasticity (and so a decreased degree of genetic determinism) also favor culture. That is, I argue that cultural transmission is not universally favored by natural selection; but the ecological conditions that do favor culture also favor phenotypic plasticity. *Homo sapiens* clearly is a cultural species and the evolutionary dynamics of a cultural species differ in important respects from that of a non-cultural species. In particular, the central tenet of sociobiology – that the behaviors that evolve are those that maximize inclusive fitness – does not hold true of a cultural species. In effect, this paper offers an evolutionary argument against the general utility of human sociobiology.

Human sociobiology, the study of the biological basis of human social behavior, has had its share of critics. Two major criticisms emerging from this barrage are: (1) Biological determinism, the view that behavior is determined by genes, is wrong when applied to humans (see, for example, Gould 1977, essay 32); and (2) Culture makes man unique in the animal world, so one cannot extrapolate from the sociobiology of non-cultured animals (e.g. the social insects) to man. The central tenet of human sociobiology is that the behaviors that evolve, and so the behaviors we observe, are those that maximize inclusive fitness. To properly understand a human behavior pattern is then to see how it maximizes fitness. Sociobiologists, especially Lumsden and Wilson (1981), have answers to the above criticisms. They argue that some animals, including humans, are flexible in their behavior precisely because such plasticity maximizes fitness. They also argue that culture is important but that it is retained on a short genetic leash – the only behaviors that evolve culturally are those that maximize fitness. Thus, they argue, the central tenet of sociobiology remains unscathed.

In this paper I shall offer what I take to be a novel account of the relation between the aforementioned criticisms of human sociobiology. After exploring the biological advantages and disadvantages of phenotypic plasticity, I shall try to explain why, and the conditions under which, culture is advantageous.

James H. Fetzer (ed.), Sociobiology and Epistemology, 57–73.

It will be shown that the central tenet of human sociobiology does not remain unscathed.

I. HERITABILITY

Abstractly characterized, evolution is the process of change in group of reproducing entities resulting form the satisfaction of the following three conditions:

(i) Variation: There is some variance among some traits of these entities.

(ii) Heritability: This variance is at least in part heritable.

(iii) Differential reproduction: Different variants reproduce at different rates.[1]

These three conditions are each necessary and (usually)[2] jointly sufficient for evolution to occur. In this section we will see how condition (ii), heritability, is relevant to the evolutionary process, and how it is independent of particular mechanisms of inheritance.

In population genetics heritability (h^2) is a highly technical notion that does not exactly correspond with its everyday counterpart. Some population geneticists *define* h^2 in genetic terms. We will consider this definition shortly, but first we need to develop a broader notion, one not committed to any specific mechanism. After defining this broader notion (following Roughgarden, 1979), we will examine its relation to the genetic notion.

A quantitative character is one that shows more or less continuous variation within a population. Examples include height, amount of pigmentation, and even such things as disposition to explore one's habitat. Any varying character that can be assigned to some metric can be considered a quantitative trait. This contrasts with characters like wrinkled versus smooth coats in peas.

What is it for a quantitative trait to be heritable in a population? Intuitively, if a trait like height is heritable then taller than average parents tend to have taller than average offspring, and shorter than average parents tend to have shorter than average offspring. This intuition forms the basis of the purely phenotypic definition of h^2 that follows.

Suppose we have some quantitative character (say number of self-grooming behaviors per day) that varies in a population. Let X_i be the value for any individual i, and $\bar{X}$ be the mean value of X in the population.

We are concerned with the relation between offspring deviation from the mean and parental deviation from the mean. Let X_0 denote the average offspring value (average number of self-grooming behaviors per day) from parents whose midparent value $((X_♀ + X_♂)/2)$ is X_p. The offspring deviation from the mean is $(X_0 - \bar{X})$ and the parental deviation is $(X_p - \bar{X})$.

A plot of offspring deviation against parental deviation (a scatter diagram) would yield a regression line whose slope is the heritability, h^2. The formula is,

$$(X_0 - \bar{X}) = h^2(X_p - \bar{X}) \tag{1}$$

(Roughgarden 1979, p. 136, formula 9.1).

Adding $\bar{X}$ to both sides yields:

$$X_0 = h^2 X_p + (1 - h^2)\bar{X} \tag{2}$$

(Roughgarden, 1979, p. 136, formula 9.2). This formula makes clear just what h^2 is. It tells us that the average offspring value in numbers of self-grooming behaviors per day depends on the midparent value to degree h^2 and on the population mean to degree $(1 - h^2)$. The values of h^2 lie between Ø and 1. If h^2 is 1 then the average offspring value is their midparent value; if Ø then the average offspring value is simply the population mean. Thus, the higher h^2 the more offspring resemble their parents in deviations from the mean.[3]

Two important points must be made about h^2: (1) As a purely phenotypic notion it says nothing about any mechanism of inheritance; (2) It is an empirical constant which must be relativized to a particular population in a particular ecological setting.

(1) h^2 is simply a measure of the degree to which parental phenotypic deviations from the mean are passed on to their offspring. The mode of inheritance is not specified and need not be genetic. For instance, in most Western societies wealth has a high degree of heritability. On the other hand, a trait with highly deterministic epigenetical development need not have a high h^2. Consider a human population fixed for a particular blue-eyed gene. There h^2 would be zero.

(2) h^2 must be relativized to a particular population and a particular ecological setting. In the above eye color example, h^2 was 0. But change the population to one where a second gene (say, one producing darker blue) is also present and h^2 would increase. In the wealth example, we could change the "ecological setting" (say, by changing inheritance laws) and lower h^2.

Although our definition of h^2 makes no commitment to any particular mechanism, in the standard cases of inheritance studied by population biologists the mechanism of inheritance is genetic. Let's turn our attention to a genetic definition of heritability.

The total variance in a phenotypic trait in a population can be divided into two types, *between-genotype-variance* and *within-genotype-variance*. Some phenotypic differences in a population are due to genotypic differences; this variance, *the between-genotype-variance*, is represented by 'V_G'. The remaining variance, obviously, is not due to genotypic differences. The name given this type of variance is *environmental variance,* V_e.[4] Thus we have the following equation (where 'V_T' stands for the total variance):[5]

$$V_T = V_G + V_e \tag{3}$$

The between-genotype variance, V_G, can be broken down into two components; the *additive genetic variance,* V_A and the *dominance variance* V_D. So

$$V_G = V_A + V_D \tag{4}$$

V_A is that part of V_G that can be attributed to the separate alleles composing the genotype (independent of their context). Thus V_A is the component which is heritable. V_D is the remaining variance.[6]

We are now in a position give a genetic counterpart to the phenotypic definition of h^2. From Equations (3) and (4), we get the following equality:

$$V_T = V_A + V_D + V_e \tag{5}$$

The additive component of the phenotype is that which is passed down genetically from parent to offspring, so of the total variance only the additive part is (genetically) heritable. Thus in cases of genetic inheritance we expect the followng quantity to be closely related to h^2:

$$\left(\begin{array}{l}\text{fraction of total}\\ \text{phenotypic variance}\\ \text{due to additive effects}\end{array}\right) = \left(\frac{V_A}{V_A + V_D + V_e}\right) \tag{6}$$

(Roughgarden 1979, p. 154, formula 9.34). In fact using standard one-locus models of population genetics it can be shown that,

$$h^2 = \left(\frac{V_A}{V_A + V_D + V_e}\right) \tag{7}$$

(see Roughgarden, 1979, pp. 155 and 156 for derivation of this formula.).

Some population geneticists take (7) as the *definition* of heritability. We cannot do so since we are concerned with non-genetic modes of heritability. Thus, we follow Roughgarden in defining h^2 as the offspring-midparent regression, and taking (7) as a result which holds only under certain conditions. As a rough approximation, one sufficing for present purposes, we will say that when (7) holds inheritance is *genetic*, and when it doesn't hold the mechanism of inheritance is *nongenetic.*[7] Thus the purely phenotypic definition of h^2 and Equation (7) give us the theoretical background necessary for drawing the distinction between biological genetic evolution and biological evolution with nongenetic transmission.

II. PHENOTYPIC PLASTICITY

Basically, phenotypic plasticity is the ability of a genotype to produce various phenotypes. Recall that the total phenotypic variance in a trait can be divided into the between-genotype-variance, V_G, and the within-genotype-variance, V_e (Equation (3) above). V_e is a measure of the range of phenotypes a given genotype can produce.

Consider a population where a quantitative character X is effected by one locus with 2 alleles. The following graph gives the phenotypic distribution functions for the three genotypes: ('$P(X/A_iA_j)$' is the phenotypic distribution

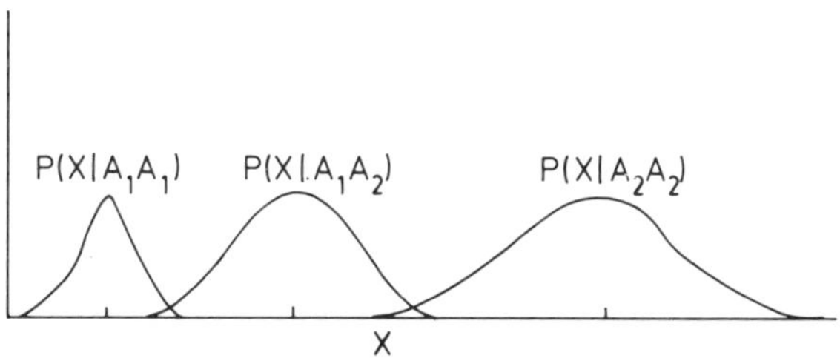

Fig. 1.

function for genotype A_iA_j.) Two points obvious from inspection are: (1) The curves for the three genotypes all have some spread; (2) A_1A_1's spread is less than A_1A_2's spread which is less than A_2A_2's spread. V_e is basically a measure of the breadth of a phenotypic distribution curve for a given genotype. The two points can be as restated as follows: (1) Each of the three curves has a nonzero V_e; and (2) $V_e(A_1A_1) < V_e(A_1A_2) < V_e(A_2A_2)$.

Put more biologically, V_e is a measure of the range of phenotypes a given genotype can produce. Roughgarden (1979, p. 216) suggests that genotypes with large V_e's be described as *phenotypically plastic*. I agree, but with this qualification: If V_e is taken as the variance of *actual* phenotypic distributions for genotypes, then V_e cannot be identified with phenotypic plasticity. Rather it can only be viewed as a measure of phenotypic plasticity, and an imperfect one at that. Phenotypic plasticity is clearly a dispositional notion. A genotype may well be able to produce diverse phenotypes in many diverse environments. But if it happens to find itself in only a small percentage of that range of environments, then its actual or manifest V_e may be rather small. Assuming the hypothesis of this example, it would be absurd to conclude that this genotype had a low level of phenotypic plasticity. Still, an imperfect measure is better than no measure at all; so, recognizing its limitations, we will take V_e as a measure of phenotypic plasticity.

III. THE SELECTIVE ADVANTAGES AND DISADVANTAGES OF PHENOTYPIC PLASTICITY

What are the selective advantages of phenotypic plasticity? In a review article on phenotypic plasticity in plants, A. D. Bradshaw (1965) suggests that high phenotypic plasticity is advantageous where fluctuations occur rapidly in time (on the order of a lifetime) or in space.[8] The reasons for this are fairly obvious. The argument becomes even more compelling when one adopts a realistic assumption about plasticity; that is, that plasticity is adaptive. By *adaptive plasticity* we mean that actual phenotypic distributions are affected by environmental demands, so that the likelihood of a given phenotype being produced is greater in environments where it is better adapted and lesser in environments where it is less adapted. For instance, in water plant arrowhead (*Saggitaria saggittifolia*) submerged leaves have a quite different shape from leaves in air. The ability to produce two shapes of leaves is plasticity. The ability to produce the right leaf in the right setting is adaptive plasticity.

We are primarily concerned with behavioral plasticity. One might think that plastic behavior is "learned behavior", that in contrast to "innate behavior". This old dichotomy is a conceptual muddle. As Mayr (1976, Chapter 47) has pointed out, behavior is better conceived of as governed by programs varying in their degree of openness. So our question is: What are the selective advantages of open-programmed behavior? The considerations discussed above seem to apply here as well. In rapidly fluctuating environments where no one behavior is optimal, it is advantageous to have the relevant behaviors governed by an open program so that appropriate behavior can be triggered by environmental cues. Here the hypothesis of adaptive plasticity is crucial.

Perhaps the major point to be gleaned from the literature on plasticity is that the more unpredictable the environment the greater the advantage of flexible, open-programmed response.[9] We have briefly covered some reasons why this is so. Roughgarden (1979, p. 216) suggests an entirely different reason why this might be so. Before presenting this suggestion we must examine the relations between phenotypic plasticity, heritability, and response to selection.

Recall our abstract characterization of evolution: evolution is the change in a population resulting from the differential reproduction of heritable variation. Why heritable? Intuitively, selection acting on nonheritable variation will result in differential reproduction, but that differential reproduction will not affect the next generation, i.e., it will have no evolutionary effect. Agreeing with that intuition, the following relation can be demonstrated for cases of directional selection:

$$\begin{pmatrix}\text{response to}\\ \text{selection}\end{pmatrix} = h^2 \begin{pmatrix}\text{strength of}\\ \text{selection}\end{pmatrix} \tag{8}$$

(Roughgarden, 1979, p. 142, formula 9.9). Thus the rate of evolutionary change depends on both the strength of selection and h^2.

From formula (7), we can see that as V_e tends to infinity, h^2 must tend to $\emptyset$ (see Roughgarden, 1979, pp. 146ff). The biological reason for this is that as V_e increases, the phenotypic distributions of the genotypes increasingly overlap. This has the effect of hiding the genotypic variation from selection, i.e. phenotypic selection is not translated into genotypic selection. Thus, the higher V_e, the lower h^2 and so the lower the response to selection.

Roughgarden (1979, p. 216) suggests that in an unpredictable environment the best evolutionary strategy[10] may be to be genetically unresponsive to environmental fluctuations. Why? In an unpredictable environment there is no discernable direction to environmental changes. Environmental states are as likely to reverse as to continue changing in a given direction. Thus the best evolutionary strategy may be to stay put genetically rather than to try to track a capricious environment. If this is so, then there would be selection for high plasticity, since as we have just stated high plasticity, high V_e, effectively shuts off genetic response to selection. (Notice that here the hypothesis of adaptive plasticity is not crucial.)

Roughgarden's suggestion may partially explain some of the observed plasticity in plant and animal species. I will argue that it has special relevance to the evolution of cultural transmission systems. But first we need to ask whether or not high plasticity is an unmixed blessing.

One might think that if the genome could "know" its environment completely and program the correct response for every environmental situation, then flexible open-programmed response would be unnecessary. Thus, in unpredictable environments high plasticity is advantageous, in predictable environments it is unnecessary but neutral. Such a view ignores the costs of learning (or, more generally, of obtaining information from the environment). This will receive further discussion later in this paper; for now let us simply note that learning takes time and energy and may often involve considerable danger.[11] (Consider for example the problem of learning who your predators are.) Like all other biological traits phenotypic plasticity is not inevitably selected for. Its selective value depends on the ecological setting and may be either positive or negative.[12]

IV. THE INHERITANCE OF ACQUIRED VARIATION – PHENOTYPIC TRANSMISSION

Thus far we have discussed some of the advantages and disadvantages of phenotypic plasticity. I will now concentrate attention on the evolutionary significance of plasticity in one particular ecological setting – a rapidly, unpredictably fluctuating environment.[13] I will argue that the optimal solution to the problems presented by such an environment involves the inheritance of acquired variation.

Recall formulas (7) and (8) above. We have seen that as V_e tends to infinity, h^2 must tend to 0, and that as h^2 goes to 0 the response to selection must also go to zero. Actually the term 'response to selection' is ambiguous. We have been using it in its evolutionary or transgenerational sense. One might also use it to refer to changes within a generation. The degree of heritability determines how much of the change within a generation is of evolutionary significance. To see just why this is so consider the following simple model.

Suppose in a given population (diploid, random mating and in Hardy-Weinberg equilibrium) height is determined by one locus at which two alleles, A_1 and A_2, occur with frequencies p and q ($p + q = 1$). Let's make this the limiting case of high V_e where the phenotypic distribution curves of all three genotypes (A_1A_1, A_1A_2 and A_2A_2) exactly coincide. Suppose we census the population at a given stage in the life cycle, call it S_1, and find phenotypes distributed normally around 5 decimeters. (see Figure 2.) The curve in Figure 2 represents the phenotypic distributions of all three genotypes prior to selection. Now suppose there is strong selection for height

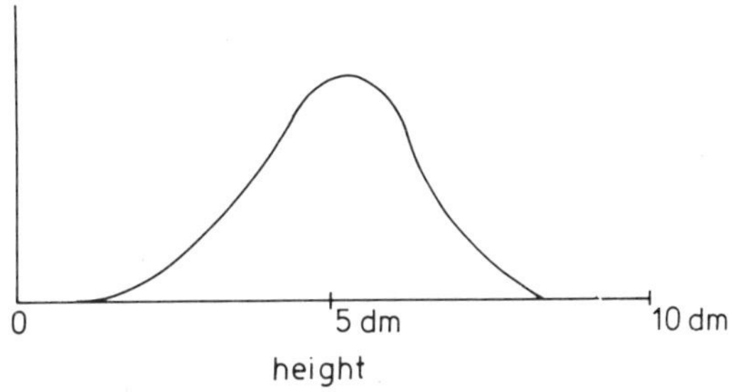

Fig. 2.

around 6.5 decimeters. This selection occurs as differential mortality just after stage S_1 and is completed prior to stage S_2. Now we census the population at S_2 and find the phenotypes distributed normally around 6.5 dm. (See Figure 3.) Again, the curve in Figure 3 represents the phenotypic distributions

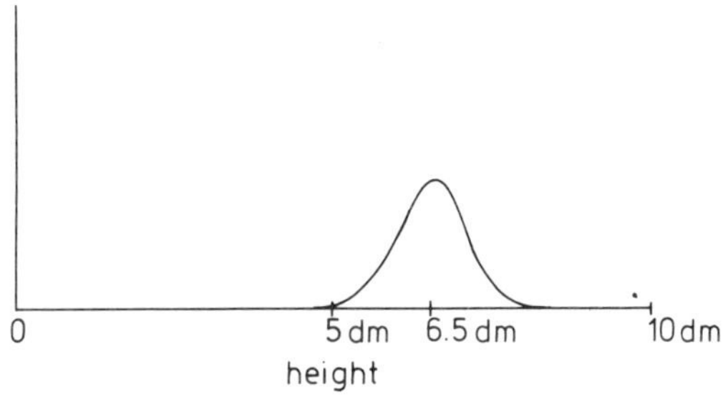

Fig. 3.

of all three genotypes at S_2. Selection has shifted the population mean. But what effect will it have on the next generation? Neither gene nor genotype frequencies have changed from S_1 to S_2 nor will they be changed by random mating and reproduction. So the curve for S_1 of the next generation will look just like Figure 2. (Excluding random drift, the only possible difference between generations is in absolute numbers of members of the population, but this is irrelevant to present concerns.) Clearly no genetic evolution has occurred; nor has any evolutionary change occurred at the phenotypic level.

The reason for the lack of change is clear in this example; it is not due to lack of variability (condition (i)) nor to lack of selection (condition (iii)), but rather to lack of heritability. Selection has not changed gene or genotype frequency, so the offspring have the same distribution of genotypes. Since phenotype distribution is a function only of the genotype, phenotype distributions also remain the same.

But suppose the phenotype probability density functions were functions of genotype *and* parental phenotypic value. For instance, suppose the offspring distribution of a $A_2A_2 \times A_2A_2$ matting with mid-parent value of 5 would differ from the offspring distribution of a $A_2A_2 \times A_2A_2$ matting with mid-parent value of 6.5. If we assume a positive influence of parental phenotypic value, the latter curve would be shifted to the right of the former. In this case we have a positive heritability, h_2, of height and Equation (8) tells us to expect an evolutionary response to selection. Let us see.

Prior to selection we again have the phenotypic distribution curve of Figure 2. Immediately after selection we again get the curve of Figure 3. But what about the next generation? That depends on the degree of h_2. Assume h_2 is 1. Then a census of the S_1 stage of generation 2 would yield the curve of Figure 3 (again there could be irrelevant differences in absolute numbers). As in the first example, selection and random mating result in no genetic changes (i.e., gene and genotype frequencies of generation 2 equal those of generation 1). But in this case we do have evolutionary change at the phenotypic level.

The above example supposes a nongenetic transmission of phenotype from parent to offspring. Shortly we will discuss possible mechanisms for this mode of inheritance (in particular, human culture) and will characterize it more precisely. But for now, let me introduce a name for it, *phenotypic transmission*.

One point that follows from the above example is that with phenotypic transmission, high heritability (h^2) – and so high evolutionary responsiveness to selection – are not incompatible with high phenotypic plasticity (V_e). This will be important in what follows.

Recall the discussion above of the main advantages of phenotypic plasticity. First, open-programmed response is advantageous in that it makes it possible for an organism to respond appropriately to environmental conditions that cannot be predicted by the genome. By hypothesis, such conditions characterize capricious environments. Second, plasticity may well have another advantage in such environments. It effectively shuts off genetic response to environmental change. By hypothesis, genetic tracking will always lag behind the environmental changes. As the changes are not directional, staying put genetically is no less successful than such genetic tracking. Furthermore, if one can assume that the capricious environmental fluctuations within environmental space are relatively circumscribed, then staying put at a genotype relatively well-adapted to the mean will be better than unsuccessfully tracking the environment. In short, from a genetic point of view, phenotypic plasticity is a better solution to the problems posed by rapidly fluctuating, unpredictable environments than is the evolution of closed programs of behavior.

However, a much better – indeed, seemingly optimal – solution exists when the right sort of phenotypic transmission mechanism is in force. We can appreciate its advantages by contrasting it with the above solution.

A feature of the above solution (*viz*. phenotypic plasticity) is that the information which fills in each organism's open program during its life is lost to the next generation. Unless there is no correlation between successive environmental states, this information will be relevant to the next generation.[14] Genetic evolution cannot gather and transmit this information for the reason already mentioned, *viz*. genetic change cannot keep pace with the environmental changes. Thus, information gathered and transmitted genetically would be consistently out of date. The advantage of the open-programmed-no-evolution strategy is that it is not misleading. It patiently awaits the input of the environment before directing the behavior of the organism. The disadvantage is that the organism has no head start.[15] That is, the organism has to expend the time and energy associated with ordinary learning. Clearly, an optimal solution would be one that was not misleading and yet gave the organism a head start.

Evolution with phenotypic transmission can be just such a solution provided the nongenetic mechanism of inheritance is of the right sort. What is needed is some mechanism capable of reliably transmitting the "wisdom" gathered in an organism's life to its descendants. The reason why this type of evolution succeeds where genetic evolution fails is that it can keep pace with a rapidly changing environment. Unlike the genetic transmission of information (which in this ecological situation would inevitably transmit misleading, out-of-date information), effective phenotypic transmission is capable of transmitting all and only that information which proved useful in the last generation.[16] Put another way, with phenotypic transmission, evolutionary change can be much faster than can genetic evolutionary change.

I argued that, in rapidly fluctuating unpredictable environments, phenotypic plasticity has a positive selective value. In this section I have argued that, if the capricious environmental changes are not *too* capricious, *i.e.*, if the benefit of receiving the still useful information from the last generation exceeds the cost of receiving the out-of-date information, then the optimal means for adapting to such environmental changes is evolution with phenotypic transmission (see Boyd & Richerson, forthcoming).

V. CULTURE

The results of the last section is of interest only if phenotypic transmission is a biologically plausible mode of inheritance. Thus far, we have characterized phenotypic transmission by exclusion; phenotypic transmission is any sort of nongenetic inheritance. That characterization is not terribly useful. A more useful characterization of phenotypic transmission comes from an automata-theoretic model of the process.

Biological organisms can be viewed as automata. An organisms's genome is the hard-wired program. The whole phenotype of the organism is the result of environmental input to this program. (See Figure 4) The organism itself

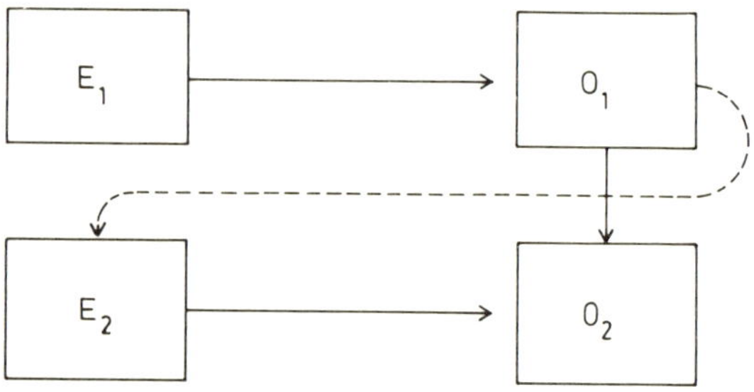

Fig. 4.

has various outputs; for our purposes we consider only two categories of outputs. The first sort has to do with reproduction. This output is a program (genome) for another organism. (Sexual reproduction is not quite so simple, but no matter.) In Figure 4 this is represented by the solid line from O_1 to O_2. This output corresponds to the mechanism of genetic inheritance. The second sort of output from O_1 also affects O_2 but not directly. Some output from O_1 modifies O_2's environment, E_2. (Of course O_1's output can also modify its own environment, but that's irrelevant here.) This sort is represented in Figure 4 by the dashed line from O_1 to E_2. Thus the eventual phenotype of O_2 is affected by both sorts of outputs from O_1.

The category of output represented by the dashed line of Figure 4 cannot be identified with phenotypic transmission. Parental effects on offspring phenotype via environmental modifications can be divided into three types: (a) those that serve to increase parent-offspring similarity with respect to a given trait or set of traits; (b) those that serve to decrease similarity with respect to the trait or set of traits; and (c) those that have no effect on similarity with respect to the trait or set of traits. Clearly, only the first of these types will count as a mode of inheritance, so it is this category that we identify with phenotypic transmission. That is, *Phenotypic transmission is any parentally-induced environmental modification which has the effect of increasing parent-offspring similarity with respect to a given trait or set of traits.*

Note that phenotypic transmission is defined in terms of *effects* of environmental modifications. No mention is made of intentionality or purpose. Perhaps the clearest cases of phenotypic transmission will involve parental teaching and offspring learning and so a strong suggestion of intentionality. That may be, but the definition does not require it.

The model of Figure 4 delimits the possible mechanisms of inheritance. They are of two sorts. The first is the direct transmission from parent to offspring of the offspring's hard-wired program. For the organisms of this planet, this is genetic inheritance. The second is the indirect transmission of information from parent to offspring via the parental modification of the offspring's environmental input. This is phenotypic transmission. There are no other possibilities. (Note that here we are concerned with possible ways for O_1 to affect O_2's phenotype, not with other possible patterns, e.g., O_1 affecting O_3 or O_2 affecting O_3.)

We have delimited the possible mechanisms of phenotypic transmission. The questions now before us are whether any such mechanisms are actualized in nature and, if so, what are they.

Various mechanisms of phenotypic transmission do exist in nature. Within *Homo sapiens* alone they range from the *in utero* transmission of heroin addiction from mother to offspring to modern human culture. A range of mechanisms is also found in some other animal species (see Bonner, 1980). We are primarily concerned with those mechanisms which could be called cultural.

In his book on the evolution of culture in animals, J. T. Bonner (1980) initially defines culture as "the transfer of information by behavioral means" (p. 10). Unfortunately this definition will not quite satisfy his or our purposes. An ant leaving a chemical odor trail from a source of food back to her

NOTES

[1] Adopted from Lewontin (1978).
[2] Differential reproduction is a force. Together with conditions (i) and (ii), it will result in evolutionary change provided it is not counterbalanced by other forces (e.g., mutation).
[3] There are important limitations on this purely phenotypic definition of h^2. For further discussion see Roughgarden (1979, chapter 9).
[4] This is not the happiest choice of terms. The idea is that the variance within a given genotype is due to environmental differences. This may well be, but one should not simply assume it. The possibility that epigenesis is an essentially stochastic process should be left open.
[5] This equation and what follows in the genetic definition of heritability require some strong assumptions. See Roughgarden (1979, chapter 9) for details.
[6] For details see Roughgarden (1979, pp. 148–51).
[7] Strictly speaking this last claim is false. In particular, in multi-loci models of genetic inheritance (7) does not hold exactly. What is true is that when inheritance is non-genetic, (7) will not be even approximately correct.
[8] Discussed in Roughgarden (1979, p. 216). See also Levins (1968) and Wimsatt (1980).
[9] For a fuller discussion see Mayr (1976, chapter 47) or Bonner (1980, especially chapter 7).
[10] Talk of evolutionary strategies is just a convenient way to refer to the results of selection. Of course the selection process involves no foresight or planning.
[11] See Mayr (1976, chapter 47) and Pulliam and Dunford (1980, especially chapter 4) for further discussion.
[12] See Brandon (1978) for an argument that no trait is invariably selected for.
[13] In this context the relevant notion of unpredictability (or capriciousness) is not absolute unpredictability but rather unpredictability-to-the-population. For a nice discussion of related matters, see Wimsatt (1980).
[14] For a more detailed discussion of the conditions under which the transmission of information gathered in one generation will be useful to the next, see Boyd and Richerson (forthcoming).
[15] Obviously, an open program gives some structure to the organism's behavior. By 'no head start', I simply mean that the organism has to wait for environmental input.
[16] By hypothesis, not all information useful to one generation will be useful to the next; but on the supposition that most of the relevant ecological changes will be non-catastrophic, some of the information will be useful.
[17] Although the transmission is behavioral, its effects need not be purely or even primarily behavioral; but presumable they frequently will be, and our focus is on behavioral traits.
[18] See Bonner (1980) for examples of culture in non-human animals.

REFERENCES

Bonner, J. T.: 1980, *The Evolution of Culture in Animals*, Princeton University Press, Princeton.

Boyd, R. and Richerson, P. J.: (forthcoming), *Cultural Evolution*, Chicago University Press, Chicago.

Bradshaw, A. D.: 1965, 'Evolutionary Significance of Phenotypic Plasticity in Plants', *Advanced Genetics* **13**, 115–155.

Brandon, R. N.: 1978, 'Adaptation and Evolutionary Theory', *Studies in History and Philosophy of Science* **9**, 181–206.

Brandon, R. N. and Hornstein, N.: (forthcoming), 'From Icons to Symbols'.

Cavalli-Sforza, L. L. and Feldman, M. W.: 1981, *A Theory of Cultural Evolution: Cultural Transmission*, Princeton University Press, Princeton.

Gould, S. J.: 1977, *Ever Since Darwin*, W. W. Norton and Company, New York.

Levins, R.: 1968, *Evolution in Changing Environments*, Princeton University Press, Princeton.

Lewontin, R. C.: 1978, 'Adaptation', *Scientific American* **239**, 212–230.

Lumsden, C. J. and Wilson, E. O.: 1981, *Genes, Mind and Culture*, Harvard University Press, Cambridge.

Mayr, E.: 1976, *Evolution and the Diversity of Life*, Harvard University Press, Cambridge.

Pulliam, H. R. and Dunford, C.: 1980, *Programmed to Learn*, Columbia University Press, New York.

Roughgarden, J.: 1979, *Theory of Population Genetics and Evolutionary Ecology: An Introduction*, MacMillan Publishing Co., New York.

Wimsatt, W.: 1980, 'Randomness and Perceived-Randomness in Evolutionary Biology', *Synthese* **43**, 287–329.

DAVID SMILLIE

SOCIOBIOLOGY AND HUMAN CULTURE

ABSTRACT. Adopting a view of macroevolution according to descriptive punctualism, this paper shows that it is possible to use developmental evidence to provide a tentative account of the evolution of human culture. Stages of language development in the human infant illustrate possible communicative strategies employed by ancestral species leading to the development of language in *Homo erectus*. It is only with this latter accomplishment that we can speak of the transmission of cultural information, a realm that requires its own ontological status.

There seems little doubt that a central challenge for sociobiological inquiry is that of providing a clear understanding of the steps by which human nature evolved over the past eight million years or so, that is, from the time of divergence between pongid and hominid phyletic lines. The task is certainly daunting, since evolutionary changes over this time period implicate a variety of biological factors that do not leave fossilized remains. We are forced to construct models of what might have taken place in behavioral realms or at the level of mental capacities and processes. Such models must be elaborated so that they bear on events and structures that leave material evidence, thus providing tests of the original assumptions. Even so, the empirical evidence is often open to a variety of interpretations and usually relates to only a small part of the proposed theoretical model. It seems to me that there is no other way to proceed with the task of understanding the evolution of human nature, and in the paper that follows I will suggest a model for the evolution of human culture which is open to these difficulties. The importance of a scientific understanding of the human species ought to compensate for the labor and uncertainty involved in the evaluation of difficult-to-test models. We simply cannot afford to wait for the maturation of the scientific disciplines required for a more objective account of our evolutionary past or of our contemporary condition.

In recent years authors concerned with human sociobiology (Cavalli-Sforza and Feldman, 1981; Durham, 1978; Lumsden and Wilson, 1981, 1983; Richerson and Boyd, 1978) have focused on human culture as the key to understanding the remarkable and peculiar adaptive patterns of modern *Homo sapiens*. Thus Lumsden and Wilson (1981, 1983) speak of "gene-culture

James H. Fetzer (ed.), Sociobiology and Epistemology, 75–95.

coevolution" to express the theme of complex feedbacks between biological individuals and an external system of socially constructed cultural entities, while other authors have postulated the existence of an autonomous pattern of cultural evolution operating independently of biological evolution. Within these sociobiological models, culture has usually been characterized loosely either as a complex sociocultural tradition or as any element or combination of elements that is transmitted from generation to generation. Unfortunately, cultural anthropologists (Kaplan and Manners, 1972; Keesing, 1974) have fundamental disagreements about the nature of culture and thus have not helped to specify the variable which is presumed to hold the key to an understanding of the evolution of human nature.

The problem in defining human culture may itself be phrased in sociobiological terms: human culture has evolved over a long period of time and through a number of different stages so that what exists today as human culture is an amalgam of different component adaptive strategies. Evolutionary change does not proceed by the sort of orderly steps of logical reasoning that we attribute to a design engineer, but rather by the utilization of bits and pieces of adaptation that have proved successful in the past. Jacob (1982) speaks of evolution as proceeding by a process of trial-and-error "tinkering" such that new features that work are adopted, old adaptive features, if they can be used at all, are retained. The organism – or, we might say, the species – is thus the product of a particular, fortuitous, evolutionary history which produces a workable adaptive compromise in the present. If we are to understand the compromise fully we will have to be able to understand the steps that went into its formation. Culture is the adaptive pattern of our own species. If we are to understand its nature we will have to reconstruct the evolutionary steps involved in its formation. It is this line of reasoning that leads me to speak here of the evolution of human culture.

Given the complexity of the topic for both anthropologists and human sociobiologists, I will propose a simplifying scheme treating only one aspect of culture here, with the understanding that other aspects will require their own, complementary, evolutionary analysis. Human culture, I shall suggest, consists of four different aspects, which can be conceptualized as four equidistant poles of an abstract tetrahedronal space. One pole designates material culture or the artifacts that are constructed and passed on from generation to generation. From the perspective of the archaeologist concerned with hominid evolution such artifacts are best represented by lithic remains, or, more broadly, by all the technological accomplishments of a given group. Tools, such as digging sticks, and tool use, are probably implicated in the transition from quadrupedal to bipedal locomotion and thus can be considered as the first manifestation of human culture. The second pole of the tetrahedron consists of culture as transmitted information about the world. A number of biologists, interested in human culture (Bonner, 1980; Cavalli-Sforza and Feldman, 1981; Richerson and Boyd, 1978), have adopted the view that culture consists of transmitted traits or entities, thereby setting up an interesting contrast between cultural information and genetic information, each of which can be understood to have evolved in its own right. (I will speak further of this conceptual pole below.)

A third pole of the tetrahedron consists of social institutions and patterns, usually presumed to be influenced in part by genetic factors and in part by social traditions. Ethnologists, concerned with the complex patterns of social organization of diverse human groups, have sometimes taken symbolically mediated social organization as a central theme for cultural analysis (Sahlins, 1976). It is useful to use the term "sociocultural system" to refer to this pole.

Finally, one may take culture to be something that is constructed by individuals, which is therefore basically psychological or mental in nature. For Goodenough (1981), culture consists in "the things one needs to know in order to meet the standards of others" (p. 50). From this perspective, culture ultimately reduces to neural circuits in the brains of individual human beings, a theme having appeal to cognitive psychologists and to those who adopt some version of methodological individualism.

To speak of these views as poles of a tetrahedron enclosing an abstract space is to allow us to speak of culture-in-general as defined by these polar limits, while a specific definition of culture might be located as an area within these boundaries. In this paper, I utilize the tetrahedron of human culture to deal solely with one particular pole, culture as transmitted information. I use the model to isolate a particular aspect of human culture for evolutionary analysis, with the understanding that the larger task is not complete until the other poles are similarly treated.

I

Evolutionary theory is concerned with processes of stasis and change over periods involving many generations. The emphasis of modern synthetic theory has been primarily on the mechanisms accounting for evolutionary change, with much less concern about the mechanisms that account for periods of stasis through evolutionary time. Darwin formulated his account of natural selection in an historical era when it was widely supposed that new species appeared suddenly through the intervention of divine influence. For Darwin, it was necessary to show how a naturalistic principle or mechanism could account for the tremendous diversity of living things without the presumption of the divine creation of new species. Natural selection was presumed to operate on inherited traits through a process of continuous selection, analogous to that of the animal breeder selecting for desirable traits in domestic animals and plants. Gradual selection through mechanisms that could be experimentally manipulated and tested became the rallying point for this new account of living nature.

Subsequent developments in evolutionary theory in the 20th century have amply confirmed the soundness of Darwin's scientific program, which has provided a broad theoretical framework for all biology. However, the assumption of gradual and continuous phyletic change has remained open to question, since the fossil record presents ample instances of long phyletic persistence (Simpson, 1967) punctuated by periods of rapid speciation. Descriptively, the fossil record – always difficult to interpret because of its incompleteness – seems to show that species of animals, such as mammals, typically retain constant species characteristics for periods of time on the order of millions of years, but also go through periods of rapid phyletic change accompanied by lineage splitting over shorter periods of time, on the order of tens of thousands of years (Stanley, 1979). This pattern of evolutionary stasis and change has been called "punctuated equilibrium" by Eldredge and Gould (1972) and has precipitated a good deal of debate among evolutionists with regard to its implications for synthetic theory (Maynard Smith, 1982; Stebbins and Ayala, 1981). It seems not unreasonable to accept the descriptive account the punctuationists have given of the tempo of evolutionary change while awaiting an explanation at the level of underlying genetic mechanisms.

The patterns of phenotypic characteristics during these periods of evolutionary stasis during which species remain relatively constant constitute what might be called "stable adaptive strategies." That is, under the influence of stabilizing selection, a species persists in an adaptive pattern involving a complex intermixture of morphological, physiological, and behavioral traits that can be understood in terms of a species strategy. Evolutionary change takes place in briefer periods during which such a strategy appears to break up, producing a variety of new forms that, when they persist, constitute an adaptive radiation. Selective pressures for evolutionary change operate primarily (but not exclusively) in periods of disequilibrium, where new ecological opportunities are realized. When we consider a phyletic lineage such as the hominids, the periods of stasis define a *chronospecies* (a species organized through evolutionary time rather than contemporaneously) that has its own particular pattern of adaptive characteristics. Evolutionary change involving marked or qualitative modification of characters requires that a phylogeny pass through several evolutionary stages. Ecological pressures resulting in evolutionary change may well differ from stage to state. When the intermediate stages do not persist as *species* in their own right, or do not leave a fossil record of the transition, the transformation is often difficult to conceptualize (Mayr, 1960).

Attempts to account for the evolution of *Homo sapiens* have often implicitly adopted a gradualist orientation, with the postulation of a single set of adaptive pressures or opportunities which might explain the transformation of a pongid ancestor into modern man. Hypotheses about the role of big game hunting, reproductive strategies, or tool use all seem reasonable as ways of accounting for hominid evolution, when we limit ourselves to the comparison of modern man with present-day chimpanzees. But none of the "prime movers" (Falk, 1980; Wilson, 1975) has stood up well under critical examination; indeed, there has been a tendency to argue that they may all be operating equally throughout the span of hominid evolution. If we adopt the descriptive account of the punctuationists, however, it would make sense to consider that different selective factors operate at different times; for example, the factors important in establishing an adaptive strategy for *H. habilis* may be quite different from those operating on *H. erectus* or *H. sapiens*.

There is fairly widespread agreement about the chronospecies in the hominid phyletic line (Pilbeam, 1984), even if there are doubts and uncertainties about how, where, and when the transitions occur. We can speak – vaguely, to be sure – about "protohominids" (or "late Miocene hominids") as existing in Africa prior to the australopithecines, roughly between eight and four million years ago, although for this span there is little extant fossil evidence besides teeth. From four million years ago until a million and a half years ago, we have evidence for a variety of australopithecine species, usually classified as *A. afarensis, A, africanus, A. boisei*, and *A. robustus*. Somewhere in this grouping exists the species ancestral to *Homo habilis*, who lived two million years ago. *H. habilis* appears to be ancestral to *H. erectus* who lived for a period of a million years or more in Africa and Asia and possibly in Europe.

Half a million years ago there began to appear modern skeletal features in the hominid record leading to the designation "archaic *H. sapiens*," a species designation that might well include Neanderthal. "Modern *H. sapiens*," or *H. sapiens sapiens*, appears fairly suddenly as a distinct species between seventy and forty thousand years ago. Thus we might summarize this rough consensus as pointing to the existence of six hominid chronospecies leading to modern man:

1. "protohominids"
2. *Australopithecus*
3. *H. habilis*
4. *H. erectus*
5. "archaic *H. sapiens*"
6. "modern *H. sapiens*"

In this paper, I shall adopt the approach that the major phyletic transformation from a pongid ancestor to modern man takes place in a series of steps, or stages, roughly consistent with this account of hominid chronospecies, over a period of eight million years or so. I adopt the view here that each of these chronospecies represents a stable adaptive pattern, consistent with the account of descriptive punctuationism. As already indicated, this view need not be understood as a call for the revision of modern synthetic theory, as in the case of Gould (1980), even though a punctuated tempo of macroevolution requires theoretical explanation.

II

The major task of those attempting to provide an evolutionary account of human nature is to show how a remarkable new kind of adaptive pattern evolved over a period of several million years. I have identified this adaptive pattern as that of human culture and have proposed that we look at a single aspect of culture, which I am calling the transmission of cultural information, while recognizing that there are further challenges in accounting for other aspects of culture such as material culture or complex sociocultural systems.

There is fairly widespread agreement among contemporary students of the problem that *Homo sapiens* differs from other animals by having developed the capacity to transmit information, not just laterally, within generations, but vertically, across generations. Our own time provides numerous examples of ways that contemporary human beings exploit bodies of knowledge about the world so as to dominate the environment. The capacity to accumulate information about the environment and then to exploit it, thus obtaining an adaptive advantage, represents the invasion of what Simpson (1953) called a new "adaptive zone" or what Whittaker *et al.*, (1973) speak of as a new "ecotopic space." That is, *Homo* appears to have utilized his capacity to exploit socially generated information about the ecology to greatly modify his relations with other species (niche variables) and the surrounding habitat. We conceptualize this capacity best by thinking of it as the social generation of a wholly new kind of resource which I will call here "cinfo" (culturally transmitted ecological information), a resource that has been barely touched, or tapped, by other species. Honey bees, *Apis melifera*, represent an interesting partial exception that can be profitably examined as a comparison case, a point to which I shall subsequently return.

When we view the adaptive strategy of *Homo*, then, as the exploitation of a novel kind of resource, cinfo, we are led to a consideration of the nature of cinfo itself, that is, the kind of entity we are dealing with, and to a concern about the nature of its transmission process, i.e., the means by which cinfo is transmitted to contemporaries as well as how it accumulates across generations providing powerful bodies of knowledge that dramatically affect the nature of the surrounding world and, indeed, the species itself. I shall raise both of these issues in this paper, starting with the latter problem which contains a seeming paradox having to do with how such a new adaptive strategy might have been initiated.

It is obvious that the problem of human language is central and we must deal with the steps by which it has evolved if we are to adequately understand the way in which cultural information is transmitted. Modern languages are exquisitely shaped as a means of transmitting cinfo, endowed as they are with design features (Hockett and Altmann, 1968) that allow for the employment of arbitrary lexical elements in an almost endless set of combinations. While psychologists and linguists sometimes speak of language as though it were itself an aspect of cinfo to be acquired and transmitted by means of a general capacity for learning, a biological perspective (Lenneberg, 1967) makes it clear that the capacity for language is an evolutionarily old system with a well established biological foundation both in the brain and in the vocal apparatus. Jerison (1975) estimates that as much as 20% of the dominant cerebral hemisphere of modern man is devoted to language processes.

From an evolutionary perspective, a paradox about the relationship between language and cinfo arises in asking how such an interaction could ever have begun. While language might be understood to have evolved as a means for the transmission of cinfo, this presumes the existence of cinfo prior to language, since its presence would constitute the raison d'être for the evolution of language. Yet how can we speak of culturally transmitted information without, at the same time, postulating a mechanism for its transmission? Language and cinfo seem to require one another and yet it is difficult to see what would explain their simultaneous appearance and evolution.

We have no examples of accumulated ecological information in animals and no examples – even among highly intelligent animals – of anything approaching a communication system evolving in response to such pressures. I have alluded above to a symbolic communication system in honey bees, the waggle dance, which allows them to transmit information about the location (distance, direction) of a food-source or of a potential hive site (Gould, 1975). But, even in this remarkable instance, there is no indication of information that accumulates across generations, no indication of a communication system that evolves in response to an endless source of knowledge about the external world. Once in place, the combination of language as a system for transmitting information and cinfo as the accumulating body of knowledge seems to initiate an autocatalytic process (Wilson, 1975), a system that leads to spiraling effects of change (Holloway, 1981). But we are still faced with the problem of understanding the initial impetus for such a process.

Since this problem is complex, I shall not try to provide a complete solution here. However, a descriptive approach to the modes of speciation described above can help in finding a solution. Mayr (1960) has discussed the difficulties of explaining the emergence of evolutionary novelties, i.e., those marked evolutionary changes that produce a qualitatively new kind of adaptive pattern furnishing the means for the invasion of new adaptive zones. Where there are no fossil remains to show how such a transition has taken place, it is often difficult to explain how these marked shifts have evolved. The feathers of birds seem ideally suited to the function of avian flight, since they are light and aerodynamically efficient. But if feathers evolved to subserve avian flight this would seem to imply that both flight and feathers evolved together. The issue is complicated by the fact that we have little as fossil evidence to show how the transition occurred. Historically problems such as these led some biologists to the conclusion that evolutionary change, in these instances, proceeded by sudden leaps where new forms happened, by chance, to be in possession of the required characters. Despite a continuing interest in the possibility of "hopeful monsters" which might explain the emergence of evolutionary novelties, the weight of evolutionary logic and evidence bears strongly against such a hypothesis.

Explanations of the evolution of major adaptive shifts are aided by the understanding that evolution progresses by stages rather than by the gradual replacement of features under a constant set of adaptive pressures. The evidence seems to suggest that reptilian ancestors of birds evolved feathers independently of their function for flight, as a system suited to insulation of a warm-blooded animal against the cold. Given the existence of feathers, we can understand how a subsequent evolutionary species might have utilized this preadaptation, for flight.

When we look back at the evolution of adaptive novelties they may appear to have evolved under constant pressures leading from an old pattern to one that subserves a new end, as though the steps were guided by an identifiable search for an adaptive breakthrough. However from a Darwinian point of view we must consider intermediate forms as representing adaptive strategies in their own right, fortuitously providing what we see, from hindsight, as a movement into a new adaptive zone. When the intermediary adaptive strategies do not survive or leave fossil remains, we are left with the impression of an unexplainable emergence of a wholly new biological type.

If we return to the relationship between language and cinfo employing principles that account for the evolution of novel adaptive strategies, then they suggest that language – or a language-like communicative system – arose prior to the existence of cinfo, subserving other functions than the transmission of ecological information. Only at subsequent stages does the communication pattern lend itself to a new role. By adopting a punctuationist view of evolutionary change, we are led to view particular chronospecies as focal points for hypothesized communicative strategies prior to the evolution of cinfo, the distinctive component of modern culture with which we are here concerned.

Since the middle of the last century the topic of the evolution of language has been in some disrepute, since it seemed to produce only idle speculation on the basis of vague or indirect evidence. And even though the topic has once again been given serious consideration (Harnad *et al.*, 1976), it is still difficult to see how we might fill in the missing pieces.

To deal with the problem of reconstructing the stages through which cinfo evolves as an adaptive strategy, we need a source of data to help us reconstruct the evolution of communicative patterns eventuating in language. I propose that these data can be extracted from the ontogenetic stages of language development through which contemporary infants and children pass. Gould (1977) has argued persuasively that there are systematic relationships between evolutionary change and ontogenetic processes: phyletic transformations often arise through modifications of developmental processes where the old sequence of ontogenetic stages is disrupted or where maturational rates are modified. Elements of old patterns are often retained in any new phyletic pattern. Relationships between ontogeny and phylogeny have been overlooked, in part owing to warranted criticism of Haeckel's "biogenetic law," which claims that the ontogenetic stages of a descendent species recapitulate in ontogeny every morphological and behavioral feature of its ancestral species. As Gould documents, such a sweeping claim has been shown to be both illogical and factually incorrect.

On the other hand, students of evolutionary change have not seriously doubted that certain features of ancestral species are retained as vestigial organs or traits in descendent species. In the embryological development of land vertebrates, an embryonic stage exists in which gill structures appear and are then modified to form parts of the structure of the head and neck. The embryonic development of the heart in land vertebrates also passes through stages that repeat structures found in fishes. We can understand the retention of these vestigial organs by recalling Jacob's notion of "evolution by tinkering." When old structures are utilized in evolution to fulfill new functions, the species retains what is workable in the old pattern even when other developmental pathways to the new solution might be more efficient. Early development, in particular, is apt to involve the retention of bits and pieces of old systems so long as they serve newly evolved morphological or behavioral patterns.

Consider the example of the grasp reflex in human neonates. During the first two or three months of life, a human infant will grasp tightly any object that stimulates the palm of its hand. Its grasp is tight and strong, enabling the infant to support its own weight briefly. Such a grasp had definite survival functions for an ancestral species, as it does in living primates today. When a monkey or ape mother moves in response to a potential predator, for instance, it moves on all fours. Infant primates in clinging to the mother's fur take an active role in effecting an escape from danger. While the grasp pattern in human infants no longer serves such a function – with mothers having no body hair and a newly evolved adaptive pattern of maternal carrying – the behavior persists, a relic of an old adaptive system, giving evidence of a particular phylogenetic history. Human patterns of reaching and grasping have been modifed to serve the ends of a bipedal animal inclined toward the manipulation of objects and the use of implements, but vestiges of the old behaviors remain.

Just as we find clues in ontogeny to old patterns of mother-infant defensive reactions, so too may we search for clues related to old systems of communication predating the development of language as it exists in humans today. What we find, of course, will represent only the possibility of an old pattern, persisting as a component part of a new system adapted for the transmission of cinfo. We must therefore use what we have in the way of evidence, building our models of evolutionary stages in such a way that we can discover opportunities for putting them to the test. There is no reason here to deny the significance of developmental evidence as a means for the reconstruction of evolutionary stages simply because we have no biogenetic law that could provide us with incontrovertible proof.

For our purposes, we may take the period between nine and thirteen months of age to represent the beginning of language development in the human infant. It is during this period that the infant begins to establish certain specific communicative signals having an identifiable functional content. Since we are investigating the possibilities of a communicative system predating language as we know it, we may speak of ur-language signals having some communicative function but not yet elements of a truly linguistic system. The signals of infants of this age are both gestural and vocal, and there is some evidence (Ricks, 1979) that the signals used by a particular infant are recognizable from one set of parents to another, even across linguistic groups. Halliday (1975), a linguist who has made an intensive study of the early language-like patterns of his infant son, finds these signals to operate in the context of social interactions such as greetings, demands, and the like. Infant communication at this age involves processes of sharing, refusing, requesting, and so on. Bates *et al.*, 1979) studied two groups of infants during this same age period, one Italian, the other American. They identified a cluster of gestures used by these infants that appeared to signal communicative intent. The gestures were highly correlated with one another and were also, as a group, positively correlated with subsequent language development. These four gestures they label "giving," "ritual requests," "communicative pointing," and "showing." The first two fit well with Halliday's account of communicative signals that mediate a process of sharing. The latter two, communicative pointing and showing, relate also to an aspect of early communication noted by Halliday, signals that draw the adult's attention to something. We might summarize the evidence from the period of infant development as suggesting the development of patterns of communicative signals that relate to interpersonal exchange with *sharing* and *showing*. These developmental findings have been corroborated by Trevarthen

(Trevarthen, 1979; Trevarthen and Hubley, 1978) using a method of detailed analysis of filmed interactions between mothers and infants. He labels the pattern of communication at this age "secondary intersubjectivity," meaning that infants focus on both objects and social partners in a pattern in which each is related to the other in a complex interactive system. This combination of attention to both object and social other follows an earlier period where either the social partner or objects represent unique foci of attention.

How can these patterns of early language-like behavior in human infants be related to presumed stages in the evolution of a communicative system prior to the use of language for transmitting cinfo? Lovejoy (1981) has argued that both factual evidence and evolutionary principles support the contention that australopithecines – four to two million years ago – had established a social pattern in marked contrast to their chimpanzee-like ancestors. He holds that these ground-dwelling hominids established a pattern of shared responsibility in the care of the young, with males foraging at a distance from a home base and bringing back food to females who had primary responsibility for the care of immatures. Such a division of labor between males and females would have allowed more intensive parental care as well as the possibility of reduced birth spacing between offspring, owing to the care provided by feamels freed from the need for long-range patterns of foraging. This picture of a division of labor between males and females, with the addition of male provisioning, would be further facilitated by the extension of tool using, established as an adaptive pattern in the late Miocene protohominids of an earlier evolutionary stage. Generalizing from archeological investigations in East Africa, Isaac (1978) arrives at a similar conclusion. Australopithecines, as yet unidentified as to species, appear to have adopted a food-sharing social strategy where males perform one kind of food-getting activity, scavenging and/or collecting of distant vegetable foods, while females perform distinctly different tasks. It seems reasonable to assume that australopithecines used a variety of tools, such as digging sticks and carrying devices, long before they perfected the patterns of stone tool manufacture which have left identifiable lithic artifactual remains.

If we were to project into this picture an account of communicative patterns for these Pliocene hominids, it would seem reasonable to suppose that they might have evolved a species-wide pattern of signals mediating the sharing of foods and implements, and possibly other objects. Tool use suggests, as well, that australopithecines might have demonstrated to each other the use or the construction of tools – how to use a carrying sling, how to sharpen a stick against a stationary rock, etc. That is, it makes sense to assume that, over and beyond those social signals commonly found in primates that mediate everyday social interactions, a system mediating acts of sharing and showing, similar to the rudimentary patterns found in present-day infants existed as well. And if such patterns existed in an ancestral species, it seems reasonable to conjecture that these patterns are repeated today as a stage in the development of language in modern human young.

From a sociobiological perspective, a communicative pattern involving reliance upon acts of sharing and showing constitutes a novel social strategy for vertebrate species. Food-sharing in contemporary chimpanzees exists only in very limited forms, as when troop members use "begging gestures" to obtain meat from a dominant male. Lovejoy (1981) and Isaac (1978) propose, by contrast, that food-sharing in australopithecines constitutes a full-blown social strategy. Similarly, while imitation exists in a number of different species (Galef, 1976), we do not find in these species communicative gestures that mediate imitation, such as specific gestures of modeling or demonstrating. Imitation exists as a response of the observer to behaviors of conspecifics but is not mediated by a set of formal communicative signals. I propose here that sharing and showing arise in the hominid phyletic line as a specific strategy, present in the australopithecines of the Pliocene.

Given the postulation of a communicative strategy for australopithecines rests largely on inference and indirect evidence, we should look with interest at the developmental period in which language-like communication arises. After a period, from two to nine months, during which infants are concerned with phatic, or non-referential, interpersonal exchanges, or with the manipulation of objects independent of the social other, we find the developmental appearance of a communicative exchange involving the sharing of objects (giving, demanding, refusing) and the demonstration of objects to others (showing, pointing). The developmental evidence supports an evolutionary argument. We can thus interpret the species-wide signals of nine- to thirteen-month-old infants as remnants of a system which has had adaptive significance in the past and which now serves as a route to the development of language-learning among our contemporaries.

III

At about 13 months of age (I speak here of averages, of course), the human infant begins a new phase of language development, the use of single worlds, largely but not entirely learned from parents, to express or communicate a new set of meanings. The child continues to use words that seem designed to elicit an adult response, words used for requesting, for demanding, or for initiating particular kinds of infant-adult interaction. However, over and beyond these communicative functions, words now appear that point out something, not as a general "this," but by providing particular names or labels to objects or events. Halliday (1975) speaks of these terms as fulfilling a "mathetic function;" they seem to serve to differentiate the world into a variety of separate things, each with its own name. For his child, all terms used in this way were spoken with a falling tone, without exception, while words requiring a social response were articulated with a rising tone. Many developmental psycholinguists have been struck by the appearance of these "pointing words" (Ricks, 1979) and by the fact that most of the words spoken during this period are nouns, and most of these name objects (Nelson, 1973).

Feldman, Goldin-Meadow and Gleitman (1978) studied deaf children growing up in families where sign language was discouraged, owing to a commitment to a particular kind of educational regimen. Although not encouraged to use signs, these children nevertheless did invent spontaneous gestural signs, when approximately as old as children who, with normal hearing, start using single words. Interestingly, the pattern of development in the spontaneous signing of these children followed, for many months, the same patterns of production employed by hearing children using vocal language. These deaf children developed a system in which they designated objects by means of points, with each point referring to a particular object rather than to objects in general. Language development by these deaf children, who were not exposed to a conventional language system by adults, confirms the claim that the developmental stages through which children pass in mastering their native tongue are a part of a universal and biologically governed system reflecting a species-typical adaptive pattern. In this particular instance the activities of deaf children also confirm the existence of a language-learning stage in which single words or gestures are employed to designate objects that are of interest to those children.

What are we to make of such a system from an evolutionary point of view? Can we presume here also that the stage of language development in a modern child can tell us something about the adaptive strategy of a Plio-Pleistocene hominid? We must first note that the infant has passed from a developmental stage in which communicative signals are used to mediate a social interaction (although, significantly, a social interaction which includes objects that are a point of concern for both communicating individuals) to a stage where there is a focus, at least in the case of pointing words, on objects themselves. We can refer to this shift in attention as an interest in, or enthusiasm for, designation.

Homo habilis, living two million years ago in Africa, is by widespread agreement (Pilbeam, 1984) the first of the hominid line that deserves the generic title "Homo." Much like australopithecines in post-cranial skeleton, he had an unusually large brain and it is this that sets him apart from other hominids living at the same time in East Africa. While the evidence is not all in, he is the first of the hominids to be clearly associated with the manufacture of stone tools. In all likelihood, he was a scavenger more than a hunter. Probably stone tools were used at butchering sites to carve up large animals and also – quite likely – to extract from the carcass animal parts to serve as implements. This picture leads to the conclusion that *H. habilis* entered into large scale cooperative activities over and beyond the patterns of food-sharing ascribed to australopithecines. The adaptive strategy for this species includes a communicative system that continues to mediate and regulate social interaction; but communicative interactions now expand to include challenges posed by complex projects in which different individuals work together while dealing with different facets of a problem. The capacity to designate objects, as such, and the ability fo differentiate between one kind of object and the next, might well have served to extend the cooperative enterprises of the australopithecines into new kinds of activities and new sorts of accomplishments.

Base camp activities probably became increasingly complex as well. The ability to indicate that one object is meant rather than another would permit efficient performance of such projects as the construction of shelters or the defense of a base camp against predators.

H. habilis, then, is the animal who names. Naming, however, both for the child between 13 and 18 months, as well as for our reconstructed *H. habilis*, is an activity that occurs in the presence of the object named, even if the name persists as a fixed designator, between one incident and the next. As such it is still a tool for dealing with the perceptually present world and does not yet have the power to speak of that which is absent or is unknown to one of the communicators. We may have an instrument that can make possible a variety of new accomplishments. We may even have a system open to endless expansion, as new names are added. But we do not yet have an instrument for the transmission and accumulation of cinfo, i.e., of information about events, objects, and situations in which the receiver of the information is not himself a participant.

IV

After eighteen months, on the average, the infant begins to combine words and to master the fundamental syntax of his native language. With the development of these new communicative tools the child is able to expand his capacity to comment on what is or has been taking place and even to deal with the imaginary in either solitary or social symbolic play. It becomes increasingly difficult to discern, through the complexity of the child's learning of a native language, the traces of past adaptive communicative patterns. However, even with this increase in complexity, we know that the child retains for a time the perspective that language bears on what is already known, reminding us that language had its beginning in a situation of shared commentary on events in which all the communicators had participated. This is one meaning of what Piaget, the Swiss psychologist/epistemologist, called "egocentricity," the belief that everything known to the child is known also to his audience.

At about two years of age, Halliday (1975) notes the first signs in the language of the child of what he calls the "informative function," that is, "the idea that language can be used as a means of communicating information to someone who does not already possess that information" (p. 21). We might conceive this ability as that of telling a story or, as Halliday puts it, "I've got something to tell you." For most of us this is so fundamental to the notion of communication that it is difficult to envision it as arising only after language learning is well under way in the child. It is this new orientation toward communication that represents the key to the evolutionary emergence of cinfo, a message that informs the receiver about a situation previously unknown to him.

If we shift from the consideration of the child to that of a eusocial insect, we may achieve a clearer perspective on this aspect of communication. The waggle dance of the honey bee consists of a series of figure eight movements performed by a forager who has located a source of nectar at some distance from the hive. Its dance, performed before an audience in the dark hive, is oriented to the vertical dimension and to the position of the sun thereby communicating to others the direction of that source. The direction of the straight run of the dance on the vertical comb specifies the direction of the source; the duration and the tempo of the dance indicate the distance. Using the terms of my argument, we might say that the honey bee forager "tells a story" about a valued resource, no matter how we choose to depict the psychological perspective of the dancer or of its audience. Von Frisch (1969) and Gould (1975) have shown convincingly that such an account holds up to careful experimental analysis, i.e., that a message is transmitted from one organism to its conspecifics about a situation that only the sender comprehends initially. This situation is remarkable in animal communication: a eusocial, invertebrate species, involved in a complex communal project, has evolved the capacity to transmit a specific type of information about a valued resource to hive mates. Further sociobiological analysis will help to provide a more detailed account of the evolution of such parallel patterns of information transmission.

Extending the perspective provided by our developmental analysis of language learning in human infants to the evolutionary story of hominid phylogeny, we are led to the conclusion that somewhere in that line a biological capacity for the transmission of information about the environment arose. It makes sense to attribute this ability to *H. erectus*, the descendent of *H. habilis*, who appeared on the scene about a million and a half years ago. *H. erectus* had a refined stone tool kit, with bifaced axes as its characteristic implement. Skeletal remains suggest that he spread out of Africa and into Eurasia, colonizing new territories and adapting to a variety of new habitats, including temperate and even arctic temperatures. *H. erectus* developed technology and sophistication that made possible the hunting of large grazing animals, demonstrating his capacity to adapt to the changing ecological patterns brought on by successive waves of glaciation during the Pleistocene. The picture is that of a highly successful and adaptable animal.

While the Acheulean tool kit of *H. erectus* remained remarkably constant over a period of a million years – so that we do not have a picture of a steadily improving body of technical information – there is reason to believe that *H. erectus* used the capacity to transmit information about the unknown to members of his group and thus to make remarkable advances beyond those of his predecessors. Lovejoy (1981) argues that the success of the hominid line rests on the strategy of male provisioning and the possibilities of intensive parental care. Here I propose that *H. erectus* makes a significant move beyond the australopithecines and *H. habilis* by exploiting the potentialities of transmitted information about distant objects and events, that is by the construction of cultural information.

V

In looking at the problem of the evolution of culture it has seemed appropriate to begin the project with an analysis of the concept "culture." I have chosen to examine human culture as transmitted cultural information, or cinfo, rather than to deal with other aspects of culture, such as sociocultural systems or material culture. The model I have developed might have been described in gradualist language, since we have found that there has been a movement from communication systems concerned with acts of sharing and showing, to communication systems dealing with the specification of objects, to a language-like system that serves to transmit new information to the auditor or receiver of the signal. However, employing developmental evidence to provide clues about the nature of these evolutionary steps, we find that each of the corresponding developmental stages points to an adaptive strategy that would have had advantages in its own right, rather than merely as a step in a movement toward something else. While we may view such a phylogenetic series as a progression toward a system serving to transmit information, these adaptive systems are perhaps best understood as equilibria that served to maintain a species-typical pattern over sustained periods of tens of thousands of generations or more.

We have arrived at an account of the ability in a species ancestral to modern man, *H. erectus*, to transmit cultural information, cinfo, from individual to individual and from generation to generation by means of a species-typical system of referential communication. Is such information best understood as a manifestation of complex social behavior, and is it thus amenable to a behavioral analysis? Does it require that we employ terms of intentionality, i.e., terms referring to the subjective experiences of the participants? Or do we find the emergence of a new kind of entity, existing over and beyond the lifetime of its originator, having its own kind of causal efficacy and its own kind of evolutionary progress?

A number of years ago Karl Popper brought together a series of essays on what he called "objective knowledge" (Popper, 1972) in which he claimed that reality requires three basic categories for its description: World 1, or physical bodies and forces; World 2, or the realm of mind; and World 3, the world of products of human minds. "World 3 may be called the world of culture. But my theory stresses the central role played by language in human culture" (Popper, 1982). Popper requires that messages transmitted by means of language be given a status in their own right as immaterial existences, coordinate with the existence of mind and with the existence of material and energic entities. What I have described as the evolutionary stages resulting in the production of cinfo in *H. erectus* is consistent with Popper's conception of World 3 entities.

The idea of immaterial entities having causal efficacy relative to both minds and the material world has not been easy for philosophers to accept, since it seems to require the postulation of kinds of ontological objects the necessity of which is difficult to prove. Biologists are prone to reduce matters of human nature to the language of behavior, which is equivalent to arguing that only World 1 objects are required for a scientific explanation. Those who uphold some version of methodological individualism might be willing to accept the view that cinfo is simply a manifestation of the experiences of individuals, translated with more or less adequacy into a story that creates another experience. But if we hold to the view that cinfo has its own reality, then we can also extend that argument to hold that human brains and human languages have evolved as agents for the transmission of ideas or stories or descriptions that are, or may be, carried from one generation to the next. Stories that serve biologically adaptive ends (stories with truth value, we might want to say, or stories with verisimilitude), after all, have selective consequences for their carriers. They survive from one biological organism to the next and select for the capacity to acquire, retain, employ, and transmit them. Once cinfo is in place, it contributes to the further consequence that brains which carry it will outlive, and outbreed, brains that forget.

This proposal carries with it an important responsibility, namely, of giving in adequate accounting of the nature and functioning of immaterial World 3 objects. It is not enough to say that cultural entities exist in their own right and that they evolve in some fashion analogous to the evolution of World 1 living organisms, a theme developed by such authors as Cavalli-Sforza and Feldman (1981), Dawkins (1976), Durham (1978), Richerson and Boyd (1978). For we have the further responsibility to indicate just what kind of existence these entities have and how they operate as selective forces on living organisms. We need an ontological analysis of cinfo, as well as a biological and a psychological analysis.

In this paper I have attempted a preliminary defense of the idea that we can give a naturalistic account of the evolution of culture in hominids. I have emphasized the evolutionary stages through which language may have passed such that cinfo is an outcome among Pleistocene *H. erectus* populations. While the model itself is open to criticism, as it should be if it has scientific status, it also points toward further problems and issues for a more comprehensive conception of human sociobiology.

New College of the University of South Florida

REFERENCES

Bates, E., Benigni, L., Bretherton, I., Camaioni, L., and Volterra, V.: 1979, *The Emergence of Symbols: Cognition and Communication in Infancy*, Academic Press, New York.

Bonner, J. T.: 1980, *The Evolution of Culture in Animals*, Princeton University Press, Princeton.

Cavalli-Sforza, L. L. and Feldman, M. W.: 1981, *Cultural Transmission and Evolution*, Princeton University Press, Princeton.

Dawkins, R.: 1976, *The Selfish Gene*, Oxford University Press, Oxford.

Durham, W. H.: 1978, 'Toward a Coevolutionary Theory of Human Biology and Culture', in A. L. Caplan (ed.), *The Sociobiology Debate*, Harper and Row, New York.

Eldredge, N. and Gould, S. J.: 1972, 'Punctuated Equilibria: An Alternative to Phyletic Gradualism', in T. J. M. Schopf (ed.), *Models in Paleobiology*, Freeman, Cooper, San Francisco.

Falk, D.: 1980, 'Hominid Brain Evolution: The Approach from Paleoneurology', *Yearbook of Physical Anthropology* **23**, 93–107.

Feldman, H., Goldin-Meadow, S., and Gleitman, L.: 1978, 'Beyond Heroditus: The Creation of Language by Linguistically Deprived Deaf Children', in A. Lock (ed.), *Action, Gesture, and Symbol: The Emergence of Language*, Academic Press, New York.

Frisch, K. von: 1969, 'Honeybees: Do They Use Direction and Distance Information Provided by Their Dancers?', *Science* **158**, 1072–1076.

Galef, B. G.: 1976, 'Social Transmission of Acquired Behavior: A Discussion of Tradition and Social Learning in Vertebrates', *Advances in the Study of Behavior* **6**, 77–100.

Goodenough, W. H.: 1981, *Culture, Language, and Society*, Benjamin/Cummings, Menlo Park, CA.

Gould, J. L.: 1975, 'Honey Bee Recruitment: The Dance-language Controversy', *Science* **189**, 685–693.

Gould, S. J.: 1977, *Ontogeny and Phylogeny*, Belknap Press of Harvard University Press, Cambridge.

Gould, S. J.: 1980, 'Is a New and General Theory of Evolution Emerging?', *Paleobiology* **6**, 119–130.

Halliday, M. A. K.: 1975, *Learning How to Mean: The Beginning of Interpersonal Behavior*, Cambridge University Press, Cambridge.

Harnad, S. R., Steklis, H. D., and Lancaster, J. (eds.): 1976, *Origins and Evolution of Language and Speech*, New York Academy of Science, New York.

Hockett, C. F. and Altmann, S. A.: 1968, 'A Note on Design Features', in T. A. Sebeok (ed.), *Animal Communication: Techniques of Study and Results of Research*, Indiana University Press, Bloomington.

Holloway, R. L.: 1981, 'Culture, Symbols, and Human Brain Evolution: A Synthesis', *Dialectical Anthropology* **5**, 287–303.

Isaac, G.: 1978, 'The Food-Sharing Behavior of Protohuman Hominids', *Scientific American* **238** (4), 90–108.

Jacob, F.: 1982, *The Possible and the Actual*, Pantheon, New York.

Jerison, H. J.: 1975, 'Fossil Evidence of the Evolution of the Human Brain', *Annual Review of Anthropology* **4**, 27–58.

Kaplan, D. and Manners, R. A.: 1972, *Culture Theory*, Prentice-Hall, Englewood Cliffs, N.J.
Keesing, R.: 1974, 'Theories of Culture', *Annual Review of Anthropology*.
Lenneberg, E. H.: 1967, *Biological Foundations of Language*, Wiley, New York.
Lovejoy, C. O.: 1981, 'The Origin of Man', *Science* **211**, 341–350.
Lumsden, C. J. and Wilson, E. O.: 1981, *Genes, Mind, and Culture*, Harvard University Press, Cambridge.
Lumsden, C. J. and Wilson, E. O.: 1983, *Promethean Fire*, Harvard University Press, Cambridge.
Maynard Smith, J. (ed.): 1982, *Evolution Now*, W. H. Freeman, San Francisco.
Mayr, E.: 1960, 'The Emergence of Evolutionary Novelties', in S. Tax (ed.), *Evolution After Darwin*. I: *The Evolution of Life: Its Origin, History and Future*, Chicago University Press, Chicago.
Nelson, K.: 1973, 'Structure and Strategy in Learning to Talk', *Monographs of the Society for Research in Child Development* **38** (1–2).
Pilbeam, D.: 1984, 'The Descent of Hominoids and Hominids', *Scientific American* **250** (3), 84–96.
Popper, K.: 1972, *Objective Knowledge: An Evolutionary Approach*, Oxford University Press, Oxford.
Popper, K. R.: 1982, 'The Place of Mind in Nature', in *Nobel Conference XVII*, Harper & Row, New York.
Richerson, P. J. and Boyd, R.: 1978, 'A Dual Inheritance Model of the Human Evolutionary Process. I: A Model', *Journal of Social and Biological Structures* **1**, 127–154.
Ricks, D.: 1979, 'Making Sense of Experience to Make Sensible Sounds', in M. Bullowa (ed.), *Before Speech: The beginning of Interpersonal Behavior*, Cambridge University Press, Cambridge.
Simpson, G. G.: 1953, *The Major Feature of Evolution*, Columbia University Press, New York.
Simpson, G. G.: 1967, *The Meaning of Evolution* (revised edition), Yale University Press, New Haven.
Sahlins, M.: 1976, *Culture and Practical Reason*, University of Chicago Press, Chicago.
Stanley, S. M.: 1979, *Macroevolution: Pattern and Process*, W. H. Freeman, San Francisco.
Stebbins, G. L.: 1982, *Darwin to DN, Molecules to Humanity*, W. H. Freeman, San Francisco.
Stebbins, G. L. and Ayala, F. J.: 1981, 'Is a New Evolutionary Synthesis Necessary?', *Science* **213**, 967–971.
Trevarthen, C.: 1979, 'Instincts for Human Understanding and for Cultural Cooperation: Their Development in Infancy', in M. von Cranach, K. Foppa, W. Lepenies, and D. Ploog (eds.), *Human Ethology*, Cambridge University Press, Cambridge.
Trevarthen, C. and Hubley, P.: 1978, 'Secondary Intersubjectivity: Confidence, Confiding, and Acts of Meaning in the First Year', in A. Lock (ed.), *Action, Gesture and Symbol: The Emergence of Language*, Academic Press, New York.
Whittaker, R. H., Levin, S. A. and Root, R. B.: 1973, 'Niche, Habitat, and Ecotope', *American Naturalist* **107**, 321–338.
Wilson, E. O.: 1975, *Sociobiology: The New Synthesis*, Harvard University Press, Cambridge.

ROGER D. MASTERS

EVOLUTIONARY BIOLOGY, HUMAN NATURE, AND KNOWLEDGE

ABSTRACT. Since major Western philosophers considered human knowledge and social values in terms of the natural science of their time, it is appropriate to assess the epistemological implications of the contemporary evolutionary theory colloquially known as "sociobiology." Recent models of natural selection indicate the need to consider human social life in terms of five distinct areas of analysis: individual development (epigenesis), within-group variation, between-group variation, the hierarchical organization of complex systems, and feedback between systemic levels. In each of these domains, modern "social biology" has striking implications for traditional philosophic questions. While some positions, like those of the Sophists or of the English Empiricists, are seriously challenged by the findings of modern evolutionary biology, Aristotelian ethics and the dialectical tradition from Plato to Hegel appear to be reinforced to a surprising degree by contemporary biological theories.

I. THE PROBLEM

Traditionally, philosophers have discussed questions of epistemology in the context of their own understanding of the latest scientific findings of their times. Since this relationship of what we today call the "natural sciences" to philosophy is often overlooked in undergraduate courses or text-books, it sometimes seems paradoxical to raise complex philosophical issues in the context of recent scientific discoveries. But just as Heisenberg among others suggested a generation ago that philosophy needed to reconsider the issues posed by modern physics (Heisenberg, 1952, 1958; Strauss, 1953; Capra, 1975), it is fitting that recent biological theories be related to human understanding. In particular, it is well to consider how the evolutionary theory known colloquially as "sociobiology" – but perhaps better described as inclusive fitness theory or social biology – might alter our understanding of human society, knowledge, and values.

Two objections that can be raised to an exploration of the epistemological consequences of contemporary evolutionary biology should be discussed at the outset, although they have been treated at length elsewhere. First, there is the charge of critics that sociobiology is vitiated, in principle, by ideological bias; second, the claim of some biologists that sociobiology will so radically change traditional theories that philosophy as well as social

James H. Fetzer (ed.), Sociobiology and Epistemology, 97–113.

science will be simply "biologicized" if not replaced by natural science (Wilson, 1975).

The charge that inclusive fitness theory (or "sociobiology") is marred by an intrinsic ideological bias has principally been made by critics who are associated with the "left" (Caplan, 1978). There is a certain irony in the fact that at least one of those who has expressed this concern (Gould, 1981) himself developed a theory of evolutionary change – "punctuated equilibria" – that has been subject to criticism as ideologically biased. There is good reason to conclude, however, that neither inclusive fitness theory nor punctuated equilibrium theory can be dismissed in this way; rather, each can be shown to be a set of scientific hypotheses whose form has in the past been associated with the most diverse political and social views (Ruse, 1979; Rosenberg, 1980; Masters, 1984b). With regard to inclusive fitness theory, for example, social behavior is derived from cost-benefit models formally akin to those used by the pre-Socratics, by Hobbes, Locke and other social contract theorists, by economists, and by game theorists; since this class of theories has been adopted by thinkers as radical as Rousseau, it cannot be dismissed on *a priori* grounds for supposedly inherent bias (Masters, 1982).

The second argument against analyzing the implications of social biology for traditional philosophy is the obverse of the first: to listen to some, one might conclude that current biological thinking renders irrelevant anything that has gone before (cf. Mayr, 1963; Wilson, 1975: chapter 1; Gould, 1977: chapter 1). This conclusion is premature and unconvincing, in good part because it seems to rest on the kind of superficial interpretation of great philosophers presented in undergraduate survey courses rather than on a serious consideration of the issues. Not only can it be shown that philosophers as diverse as the ancient Greeks (Masters, 1978), Rousseau (Masters, 1979), and Kant (Hjort, 1938) directly addressed the same themes as contemporary biologists, but a philosophical perspective can enrich the understanding of issues raised by evolutionary biology (Masters, 1983a).

In short, there is every reason to believe that recent theories in the life sciences – like modern physics – will continue to be relevant to our understanding of human nature and its place in the universe. But to explore the extension of evolutionary theory to human thought and action, one must avoid oversimplification. Not only would it be an error to equate a scientific theory with one or a few of the leading *scientists* associated with it, but we should beware of oversimplifying the basic concepts and their application. Especially as applied to human society, contemporary models of animal social behavior are far more complex than would appear from popularized discussions of "sociobiology." Indeed, I will argue that no less than five distinct levels of analysis must be considered in any exploration of the epistemological implications of sociobiology.

II. THE CHARACTER OF SOCIAL BIOLOGY

In the broadest sense, "social biology" (an older and more neutral term I will use to refer to what is conventionally called "sociobiology") represents a fundamental contribution to the growing body of theory and empirical research linking the biological and social sciences. In different ways, biologists like William Hamilton, Richard Alexander, Edward O. Wilson, John Maynard-Smith, Richard Dawkins, and David Barash have made a major step toward the analysis of human social systems from the perspective of evolutionary biology. I will assume familiarity within the main points of this approach, which studies animal social behavior in terms of inclusive fitness and has been extended by some to the process of "gene-culture coevolution" in humans (for a good textbook introduction, see Barash, 1977; for the best treatment relating the theory to human behavior, see Alexander, 1979).

Here, it can be assumed that social biology is the attempt to integrate individual development, social processes, and evolutionary trends into a single evolutionary framework. This implies that, in general, natural selection should be among the forces shaping individual and group behavior in humans as in other species. Insofar as this is true, it follows that at least some features of human thought and behavior should be related to the same forces of selection which shape the behavior and morphology of other animals, favoring traits that enhance differential "reproductive success" over the long run.

In simple language, the essence of this approach can be restated in terms of five fundamental concepts: *individual development* (or "epigenesis"), *within-group variation, between-group variation, hierarchy* (or the irreducibility of different levels of organization in living systems), and *feedback*. While it is presumed that each of these five concepts refers to phenomena that can be analyzed in terms of the neo-Darwinian or "synthetic" theory of evolution, the implications of a biological approach to social behavior are more accessible if each is considered separately.

1. *Individual Development* (Epigenesis)

In contrast to the persistent habit of distinguishing between genetic and environmental "causes," biologists today stress *interactions* between the

as well as social circumstances, constrained but not narrowly limited by the phylogenetic inheritance (or developmental "biogram") of the species.

In contrast to the conventional alternatives, the perspective of social biology leads to a paradoxical position: although a wide variety of socio-cultural norms naturally emerge in response to varied environmental conditions, a narrower range of these norms is in some sense closer to the species-typical "biogram" that characterizes what we call, in colloquial speech, "human nature." Such an approach appears complex and often satisfies no one. Relativists and historical determinists in the traditions of Thrasymachus or Marx would object to *any* notion of a "naturalistic" foundation of ethics; those who believe in divine or natural law would object to the acceptance of different standards as acceptable, depending on time and place. But the intermediate position derived from social biology does have an old and honorable grounding, since it approximates very closely Aristotle's teaching – particularly as set forth in the *Nicomachean Ethics*, V.vii (see Masters, 1975, 1978 and the following section).

4. *Hierarchy*

The relationship between individual developmental processes, within-group variation, and between-group variation can and must be "hierarchical" in nature. As with many other problems in both logic and empirical science, such a hierarchical approach often untangles what otherwise seemed intractible controversies (Anderson, 1972; Pattee, 1973). In particular, from this perspective, it is neither necessary nor even possible to seek narrowly reductionist or deterministic "causes" for complex social phenomena.

One example may be helpful here. It has been argued that natural selection operates to shape social behavior as well as physical structures, and their among the pressures in this direction, "kin selection" – and hence the tendency of animals to cooperate preferentially with close kin – will be particularly likely to evolve. But it is generally assumed that mammals have not developed a special sensory apparatus permitting the direct detection of genetic similarity in conspecifics; the *mechanisms* (or "proximate causes") of cooperation with close kin rather than unrelated animals are presumed to be such factors as prolonged co-residence, phenotypic similarity, or individual recognition by odor and sight. These mechanisms could presumably evolve as effective criteria for social cooperation because they would normally be so highly correlated with close kinship under most conditions that natural selection would be unlikely to favor the detection of more complex cues concerning degrees of relatedness.

The hierarchical character of evolutionary theory thus points to the separate operation of selective processes at the levels of the sensory modalities of the species, the phenotypical development of the individual organism, the organisation and behavior of the group, and the phylogenetic change of the species; clearly ecological systems have properties that are not entirely deducible from the behavorial repertoire of a single species in a food chain. Nature is complex. Even where simple models are useful in specifying likely social behavior at one level of analysis, such heuristics typically cannot explain other phenomenological levels (Masters, 1981).

It follows that simplistic dichotomies are unlikely to apply in a consistent fashion across broad ranges of phenomena. In classical philosophy, Aristotle provides an excellent illustration of the way an appreciation of such complexity leads to epistemological accuracy: whereas the Sophists had stressed a sharp distinction between "nature" and "convention," Aristotle emphasized the subtler interactions between natural and individually learned or culturally variable factors in human behavior (Masters, 1978). It may be useful to introduce epistemological questions to College Freshmen in terms of the "nature-nurture debate," but serious thought based on contemporary natural science needs to avoid such formulations.

While the recognition of complexity can contribute to scientific clarity, it also helps to illuminate the issues of political and moral philosophy. The Aristotelian concept of that which is "just" or "right" "according to nature" is intermediate between moral relativism and absolutism. To use the terms introduced above, Aristotle's position considers the between-group variations in our species as "natural" adaptations to environmental conditions, yet at the same time emphasizes the limits to the within-group variation consistent with "normal" or "natural" individual development ("human nature"). In other species, evolved traits at one level become constraints or givens at another; as applied to human beings, this means that individual development or "epigenesis" exhibits a species-specific or "natural" pattern that can be relevant to the judgment of varied social norms – even though these norms are, at the group-level, varying "natural" adaptations to the environment.

An example may help. At the level of individual development, human infants must be nurtured through a long period of dependence due to the phenomenon often called "neoteny" (for a fuller treatment, see Gould, 1977). As our species has evolved, some form of parental care is therefore necessary for normal maturation – and, indeed, one can point to specific responses, such as maternal imprinting on the newborn's face or cry, which presumably were selected to facilitate such care (von Cranach *et al.*, 1979). But even if one were to conclude that parental care is in some sense of the word "natural" on an individual basis, clearly there is within-group variation in parenting ability in chimpanzees as well as in humans; moreover, infanticide – the definitive demonstration of a failure to provide prenatal care – has been observed among both primates and humans in those social settings where it can probably be explained by the effect of ecological pressures and social stratification on mating strategies (Gruter and Bohannan, 1983; Masters, 1984a).

It would be simply false to assert that infanticide is "unnatural" *because* parental care is "natural" in humans as well as other primates. To argue in this way confuses distinct levels of analysis: parental care is a central feature of the normal primate and human pattern of epigenesis, but in some environments, social behaviors diverging sharply from such "species-typical" parental care may become adaptive strategies for some individuals. In the terms used here, "between-group variation" can include substantial differences in the range of "within-group" reproductive strategies, including some individual behaviors that diverge sharply from what appear to be the species-typical or central tendencies of normal development. Hence, precisely *because* social biology can contribute to a clearly understanding of the "naturalistic" foundations of social and moral judgment, it is prudent to *refrain* from speaking of specific behaviors and social institutions in absolute terms as "natural" in all times and all places. As Aristotle puts it:

> Of political justice part is natural, part legal [or culturally variable] . . . while with us there is something that is just even by nature, yet all of it is changeable; but still some is by nature, some not by nature (*Ethics*, V.vii.1134b).

5. *Feedback*

While biological properties of specific organs, individuals, social groups, species, and ecosystems are distinct, clearly each level has effects on the others. Often, we speak of "feedback" to describe the long-term causal relations between a "lower" level or more dependent system and phenomena that appear to be of a "higher" order. In such instances, the temporal or spatial context of analysis can change the way biologists interpret the variables described as "causes" and "effects."

In no area is this more evident than in the effects of behavior on evolutionary processes. It is frequently claimed that social biology can "explain" animal behavior in terms of the principles of natural selection. But, as Simpson pointed out a generation ago (1958), behavioral changes are themselves a major agent in evolutionary change. Sensory discriminations, behavioral patterns, and social structures that have been produced by ecological pressures can in turn lead to changes in the environment; once a social structure emerges as the result of individual behavior, the organism's environment includes both a physical or ecological setting and the group structure. While obviously important for other species, this type of "feedback" is particularly important in human evolution.

For example, human societies have different degrees of social stratification. Although it must be said that we still lack agreement on the origin of political institutions like the State, to say nothing of more complex forms, it is entirely possible that evolutionary theory will provide a comprehensive explanation of the origins of the different social systems observed in human cultures (Masters, 1983a). Even so, however, the individual within any culture faces situations which are defined by the existing social structure and physical environment (Masters, 1983b, 1984a); hence, at the individual level, cultural norms and social organizations would appear as environmental circumstances rather than as human creations easily subject to manipulation.

From this perspective, the tendency to speak of "Man" in the abstract is particularly dangerous, since it leads to a confusion between the behavioral potentialities of the individual human being, the group, and the species. Due to the effects of the hierarchical organization of living systems and the "feedback" between levels, an attribute which appears to be evidence of plasticity at one level may have contrary effects on other levels. For example, the saliency of individual learning during the developmental process of humans is often described as an element of plasticity (as indeed it is at the level of the individual); at the level of the group, however, socialization processes function to produce *conformity* (not to mention such behaviors as stereotyping, xenophobia, and prejudice).

In epistemological terms, it follows that the same concept can readily apply to very different phenomenological or causal relationships, depending on the context. The example of infanticide, given above, should suffice as a reminder that what is "natural" for the species-typical pattern of development need not be "natural" for a particular individual in a particular group (even among chimpanzees, not to mention humans). Traditionally, the philosophers who have been most sensitive to these properties of existence have been dialecticians: from Plato to Hegel, some thinkers have excelled at teaching us the need to respect the transformation of meanings in varied contextual settings. Unlike relativism, dialectical thought seeks the underlying structural rules producing such transformations (and frequently, the inversions) in the phenomena observed by any thought process claiming scientific status.

All too often, non-specialists dismiss a philosopher like Plato because his works are understood in a superficial, undialectical way; actually, if approached from the perspective of the principles characterizing modern science, there is every reason to respect the Platonic dialectic as an approach to the underlying processes of being and knowing (Heisenberg, 1958; Masters, 1978). Similarly, although Hegel appears to the average reader as impenetrable and confusing, a study of modern political systems from the perspective of social biology leads to conclusions that are closer to Hegelian thought than to either Locke or Marx (Masters, 1975, 1983b).

III. CONCLUSION

Traditional philosophers raised questions of human knowledge in the context of a serious attempt to link what we today call natural science with an understanding of the human condition. In contemporary universities, the division between "natural science," "social science," and the so-called "humanities" has obscured the fundamental linkages between disciplines that are necessary to approach these issues. What is called "epistemology" cannot fail to be illuminated, therefore, by a study of contemporary evolutionary biology – and therewith, by a deeper understanding of the nature of human perception and thought as well as of the character of human social life and culture.

It is obvious, for example, that such fields as the neurophysiology of perception, linguistics, and artificial intelligence can illuminate the process of human knowledge. But insofar as natural science shows the relationship between the sensory apparatus of an animal, its behavioral repertoire, and the resulting structure of social groups, it is hard to see how one could exclude evolutionary theory from the range of disciplinary approaches relevant to epistemology. Quite the contrary, as I have argued above: the more exact our analysis of biological science, the more directly relevant it will appear to such questions.

Conventionally, a field like social biology is either dismissed by humanists and social scientists on such grounds as "genetic determinism" or "ideological bias," or it is proclaimed by scientists as a substitute for traditional philosophical understanding. Perhaps over the next generation we shall find that neither response does justice to the continuities between contemporary natural science and traditional philosophy. As evolutionary theory since Darwin should teach us, change is not automatically "progress" in the sense of linear improvement: whether in the evolution of species or in the history of thought, perhaps the time has come to return to a deeper respect for the underlying principles constituting the nature of things.

Dartmouth College

REFERENCES

Alexander, Richard D.: 1979, *Darwinism and Human Affairs*, University of Washington Press, Seattle.

Antiphon: 1923, *Discours, Suivis des Fragments d'Antiphon le Sophiste* (ed. Louis Gernet), Editions des Belles Lettres, Paris.

Anderson, P. W.: 1972, 'More is Different', *Science* **177**, 393–96.

Axelrod, Robert: 1984, *The Evolution of Cooperation*, Basic Books, New York.

Axelrod, Robert and Hamilton, William D.: 1981, 'The Evolution of Cooperation', *Science* **211**, 1390–96.

Barash, David: 1977, *Sociobiology and Behavior*, Elsevier, New York.

Caplan, Arthur (ed.): 1978, *The Sociobiology Debate*, Harper & Row, New York.

Capra, Fritjof: 1975, *The Tao of Physics*, Shambala, Berkeley.

Cavalli-Sforza, Luigi L., and Feldman, M. W.: 1981, *Cultural Transmission: A Quantitative Approach*, Princeton University Press, Princeton.

Chase, Ivan: 1980, 'Cooperative and Noncooperative Behavior in Animals', *American Naturalist* **115**, 827–57.

Dobzhansky, Theodosius: 1966, *Evolution, Genetics, and Man*, Wiley, New York.

Durham, William: 1979, 'Toward a Coevolutionary Theory of Human Biology and Culture', in Napoleon Chagnon and William Irons (eds.), *Evolutionary Biology and Human Social Behavior*, Duxbury Press, North Scituate, Mass., pp. 39–59.

Gould, Stephen J.: 1977, *Ontogeny and Phylogeny*, Harvard University Press, Cambridge.

Gould, Stephen J.: 1981, *The Mismeasure of Man*, Harvard University Press, Cambridge.

Gruter, Margaret and Bohannan, Paul (eds.): 1983, *Law, Biology, and Culture*, Ross Erikson, Santa Barbara.

Hamilton, William: 1964, 'The Genetical Evolution of Social Behavior', *Journal of Theoretical Biology* 7, 1–52.

Heisenberg, Werner: 1952, *Philosophic Problems of Nuclear Science*, Faucett, New York.

Heisenberg, Werner: 1958, *Physics and Philosophy*, Harper & Row, New York.

Hirschleifer, Jack: 1980, 'Evolutionary Models in Economics and Law: Cooperation versus Conflict Strategies', UCLA Department of Economics Working Paper 170.

Hjort, Johann: 1938, *The Human Value of Biology*, Harvard University Press, Cambridge.

Kuo, Zing-Yang: 1967, *Dynamics of Behavior Development*, Random House, New York.
Lorenz, Konrad: 1965, *Evolution and Modification of Behavior*, University of Chicago Press, Chicago.
Lorenz, Konrad: 1970–71, *Studies in Animal and Human Behaviour*, 2 vols., Harvard University Press, Cambridge.
Lumsden, Charles and Wilson, Edward O.: 1981, *Genes, Mind, and Culture*, Harvard University Press, Cambridge.
Margolis, Howard: 1981, *Selfishness, Altruism, and Rationality*, Cambridge University Press, Cambridge, England.
Masters, Roger D.: 1970, 'Genes, Language, and Evolution', *Semiotica* **2**, 295–320.
Masters, Roger D.: 1975, 'Politics as a Biological Phenomenon', *Social Science Information* **14**, 7–63.
Masters, Roger D.: 1978, 'Classical Political Philosophy and Contemporary Biology', Paper presented to Conference for the Study of Political Thought, Chicago, Ill. (*Politikos*, in press).
Masters, Roger D.: 1979, 'Jean-Jacques is Alive and Well: Rousseau and Contemporary Sociobiology', *Daedalus* (Summer), 93–105.
Masters, Roger D.: 1980, 'Hobbes and Locke', in R. Fitzgerald (ed.), *Comparing Political Thinkers*, Pergamon, New York, pp. 116–40.
Masters, Roger D.: 1981, 'The Value – and Limits – of Sociobiology', in Elliot White (ed.), *Sociobiology and Human Politics*, Lexington Books, Lexington, Mass., pp. 135–65.
Masters, Roger D.: 1982, 'Is Sociobiology Reactionary? The Political Implications of Inclusive Fitness Theory', *Quarterly Review of Biology* **57**, 275–92.
Masters, Roger D.: 1983a, 'The Biological Nature of the State', *World Politics* **25**, 161–93.
Masters, Roger D.: 1983b, 'Social Biology and the Welfare State', in Richard F. Tomasson (ed.), *Comparative Social Research, 1983*, JAI Press, Greenwich, Conn.
Masters, Roger D.: 1984a, 'Explaining "Male Chauvinism" and "Feminism": Cultural Differences in Male and Female Reproductive Strategies', in Meredith Watts (ed.), *Biopolitics and Gender*, Haworth Press, New York, pp. 165–210.
Masters, Roger D.: 1984b, 'Evolution and History in Political Thought from Aristotle to Marx', Paper presented to Symposium on 'Punctuated Equibria vs. Gradualism: Political Implications', Annual Meeting of American Association for the Advancement of Science, New York, May 25, 1984.
Maynard-Smith, John: 1978, 'The Evoluton of Behavior', *Scientific American* **239**, 176–92.
Mayr, Ernst: 1963, *Animal Species and Evolution*, Harvard University Press, Cambridge.
Pattee, H. H. (ed.): 1973, *Hierarchy Theory*, Braziller, New York.
Rosenberg, Alexander: 1980, *Sociobiology and the Preemption of Social Science*, Johns Hopkins Press, Baltimore.
Ruse, Michael: 1979, *Sociobiology: Sense or Nonsense*? D. Reidel, Boston.
Simpson, George G.: 1958, 'The Study of Evolution: Methods and Present Status of Theory', in A. Roe and G. G. Simpson (eds.), *Behavior and Evolution*, Yale University Press, New Haven, pp. 7–26.
Stent, Gunther: 1972, 'Cellular Communication', *Scientific American* **227**, 49–51.
Strauss, Leo: 1953, *Natural Right and History*, University of Chicago Press, Chicago.

Tiersky, Ronald: 1980, 'Confusion and Uncertainty in Comparative Politics', Johns Hopkins University Bologna Center, Occasional paper 30.

Trivers, Robert: 1974, 'Parent-Offspring Conflict', *American Zoologist* **14**, 249–64.

von Cranach, Mario (ed.): 1976, *Methods of Inference from Animal to Human Behavior*, Mouton., The Hague.

von Cranach, Mario *et al.* (ed.): 1979, *Human Ethology: Claims and Limits of a Discipline*, Cambridge University Press, Cambridge, England.

Wheelwright, Philip (ed.): 1966, *The Presocratics*, Odyssey, New York.

Whiting, J. W. M.: 1969, 'Effects of Climate on Certain Cultural Practices', in A. P. Vayda (ed.), *Environment and Cultural Behavior*, Natural History Press, Garden City, N.Y., pp. 416–55.

Wilson, E. O.: 1975, *Sociobiology: The New Synthesis*, Harvard University Press, Cambridge.

LAURENCE THOMAS

LOVE AND MORALITY: THE POSSIBILITY OF ALTRUISM

ABSTRACT. I endeavor to show that biology is more of an ally to morality than one might be inclined to suppose. My strategy is to look at the significance of love in the context of evolutionary theory and then make some careful extrapolations. By distinguishing between what I call transparent love and opaque love, I attempt to show that evolutionary theory favors our having the moral attitude of minimal respect for others. I also bring out the significance which love plays in our acquiring the virtue of self-command. I conclude with a few ruminations comparing my approach with the approaches of Thomas Nagel and of Philippa Foot.

A characteristic of morality is that it calls for a measure of altruism (theories of ethical egoism aside). For from time to time we are asked to set aside, at least temporarily, our interests for the sake of others; and kindness is thought to be a virtue. But as evolutionary biology is often presented,[1] it would seem that there is precious little room for genuine altruism as opposed to what appears to be altruism. If, in the end, parental love and everything else is to be explained in terms of genes maximizing their numbers in the gene pool,[2] then what room is there for genuine altruism on our part?[3] My aim in this essay is to show that evolutionary biology is far more of an ally to morality than, at first glance, one would be inclined to suppose.[4] My strategy shall be to look at the significance of love in the context of evolutionary theory and then make some careful extrapolations. Key to the argument shall be the distinction that I shall introduce between transparent and opaque love.[5] In the final section of this essay, I shall offer a few ruminations comparing the approach taken here with the approaches of Thomas Nagel and of Philippa Foot.

Now, I realize all too well that, for some, evolutionary biology is an ally they would just as soon do without, given their concern to avoid the realm of biological determinism. Crudely put, this is the view that most, if not all, aspects of human behavior, including social interaction and various wants and desires, are controlled or, at any rate, strongly influenced by our genes.[6] Let us think of this as thorough-going biological determinism. While I cannot here become enmeshed in this controversy, let me say this. There can be no doubt that some aspects of human behavior are at least strongly influenced by our genes. However, from this it would not seem to follow that we are

James H. Fetzer (ed.), Sociobiology and Epistemology, 115–129.

II

Now, from the moral point of view, what makes parental love, as an instance of transparent love, of tremendous interest is that the attitude of respect towards others that morality requires of us is not unlike parental love with respect to transparency: *Whether or not we should (fail to) find ourselves attracted to others on account of their attributes*, there is a minimum amount of respect which all persons are owed simply in virtue of being such. An account of this right is not formulated with reference to proper names or indexicals. If all persons have certain rights (on a rights-based theory) or are owed certain duties (on a duty-based theory), then they should be treated accordingly regardless of their attributes. Generality and universalizability are regarded as formal constraints of the right.[12]

Parental love and the moral attitude of respect for persons are both referentially transparent, therefore, since neither is grounded in the attributes of the person. I do not deny that the proper way of treating people can ever have something to do with their attributes. My point, instead, is that the minimum respect which morality requires us to have toward all persons is not in any way tied to specific attributes of persons, to wit: skills, physical features, and traits of character, personality, and style. At the very minimum, one might suppose that morality requires that we treat all persons fairly and no one cruelly – and that we should do so in spite of how a person may fare with respect to the aforementioned attributes. We would not know what to make of a person who supposes that he has a moral reason to treat so-and-so (a person) fairly, since so-and-so possesses X rather than Y, where each stands for a different skill, physical feature, or trait of character, personality, or style.

The relevance of biology to all of this is as follows. As we saw earlier, if there is any capacity which is essential to the survival of the human species, it is the capacity for parental love. However, since what is characteristic of this love is that it is referentially transparent, what follows, then, is that the capacity for transparent love is essential to the survival of the human species. If, as we may reasonably assume, the capacity for parental love has been selected for,[13] then so has the capacity for transparent love; for as I have unpacked the latter concept, parental love is an instance thereof. There is a conceptual point about the nature of parental love. But, now, the capacity for transparent love entails the capacity to be favorably disposed toward someone in the referentially transparent mode. And if my characterization of the moral point of view is sound, then we have seen that it is this latter capacity which morality also requires. For, as I have argued, the idea that we should have a minimal amount of respect toward all persons makes no reference to proper names, indexicals, or attributes which persons are capable of having.

Of course, whether we are talking about transparent or opaque love, if we love a person, then we will identify with her or his interests to a far greater extent than morality requires us to do. For instance, the person's flourishing (hence, the person's setbacks) will matter to us to a far greater extent than morality will require it to matter to us. So, the favorable stance towards others that morality requires of us is not of the same strength as the favorable stance that we have towards those whom we love.[14] So the disposition to be favorably disposed toward someone in the referentially transparent mode admits of a continuum, with respect to identifying with a person's good. The favorable disposition which parental love types is at one end, whereas the favorable one which morality types is at some distance from that end, though not in fact at the other end. Lest there be any misunderstanding, I should emphasize the fact that the continuum here is with respect to identifying with a person's good. The favorable disposition of morality and the favorable disposition of parental love are equally transparent, though they involve different degrees of identifying with the good of others.

Perhaps an example at this point will prove helpful. Consider the difference between my doing for a perfect stranger whom I will never encounter again what morality minimally requires of me and my caring for my just-born infant. The fact that the attributes of neither the stranger nor the child explain my behavior is no bar to my identifying with the good of the latter more than with the good of the former. I do not deny that there is a difference between the stranger and my child. What I deny, rather, is that it is the attributes of either which carries the day in explaining my behavior. True enough, that the child is mine is of no small significance; however, as we have seen, the child's being mine is not an attribute of her or him.

Parental love involves the capacity to value someone without regard to their attributes. Morality involves this very same capacity to a much lesser extent. Respectively, we have the capacity to be very strongly favorably disposed in the referentially transparent mode and the capacity to be weakly favorably disposed in that mode. This constitutes (part of) a continuum. Now, might we not suppose that this (much of the) continuum comes in the wake of the capacity for parental love and, therefore, that if human beings (or, for that matter, any creatures) are biologically endowed with the former capacity, then they are biologically endowed with the latter, unless the latter is specifically selected against? That we have a continuum here is unquestionable; the only thing that can be at issue is whether or not it – the continuum, and, in particular, the latter capacity – comes in the wake of the capacity for parental love. And I assume that this latter capacity does unless it is specifically selected against.

From the evolutionary standpoint, the primary reason for thinking that among thuman beings the capacity for a weak favorable disposition in the transparent mode has not been selected against is that social cooperation plays a preeminent role in the survival or, at any rate, the betterment of the human species. This is unmistakably the lesson to be learned from social contract theory. All such theorists argue that in the long run people are better off for cooperating with one another. From the standpoint of parental investment, which is generally regarded as one of the cornerstones of evolutionary biology, it can easily be seen why cooperation has considerable dividends for the human species.

As I remarked at the outset, the natural parent-offspring relationship among *Homo sapiens* is the longest and most complex amongst any species. This means that raising offspring to the point where they are able to leave behind progeny is a matter of considerable parental investment for human beings.[15] Indeed, a great deal of time and energy is involved. The amount of care and attention which human offspring require is so great that humans would hardly be able to provide these things, if the possibility for social cooperation amongst human beings did not exist. It is not just a morally good thing that human beings can live in harmony and, hence, engage in social cooperation withone another. The amount of care and attention that human offspring require necessitates that humans can live this way, at least to some extent.

Now, to be sure, the fact that human beings have the capacity for social cooperation hardly entails that they have the capacity to be weakly favorably disposed towards persons in the referentially transparent mode. However, this latter capacity clearly strengthens the former. And this fact, when taken in conjunction with the above remarks, certainly supports the conclusion that the capacity to be weakly favorably disposed towards persons in the referentially transparent mode would not be selected against. In fact, given how indispensable social cooperation is, if human beings are to be able to devote sufficient time, care and attention to their offspring, it is not implausible to suppose that the capacity to be weakly favorably disposed in the referentially transparent mode would have been selected for if it did not come in the wake of the capacity to be strongly favorably disposed in that mode.

If the foregoing discussion is sound, then there is no reason to suppose that evolutionary biology is hostile to morality and, therefore, that being moral is a matter of going against the biological grain. On the contrary, there is a desirable congruence between the two, since our having the wherewithal to adopt the moral point of view, that is, our having the capacity to act morally, is favored by evolutionary biology.

Now, it may be objected that while perhaps I have shown that there is a congruence between morality and biology, I have nonetheless failed to show that this is so because evolutionary biology allows for genuine altruism if, in the final analysis, our being capable of the minimal attitude of respect, which morality requires, is to be explained in terms of natural selection.

This objection, however, is not well-taken. It confuses the content of a desire with its origins. Let us say, for the sake of argument, that we have altruistic behavior on the part of *A* if and only if *A* acts out of the desire to benefit another and with no concern for his (*A*'s) own good. We can imagine *A* coming to have this desire in any number of ways, without the genuineness of the desire being called into question. God or a great brain surgeon could have implanted it there. Or, it could be the result of a genetic mutation. Likewise, from the fact, if it is one, that the capacities for parental love and the minimum attitude of respect which morality requires can be explained by way of natural selection, it in no way follows that the desires which are constitutive of those capacities cannot be altruistic.

III

Thus far I have argued for the possibility of morality by endeavoring to show that the attitude of respect which morality requires of us – and so the kind of favorable referentially transparent disposition which it requires – comes in the wake of parental love. Though still concerned with showing the relevance of love to morality, I now wish to argue for the possibility of morality from a somewhat different direction. It is obvious that from time to time morality requires that, out of respect for others, we restrict our behavior in various ways. So restricting our behavior can either be seen as a burden or as a deep expression of something that we desire to do. When it is the latter, I shall regard so restricting our behavior as a form of self-command. In what follows I want to suggest how the way can be paved for restricting our behavior to be an instance of the latter.

On at least a very crude Freudian account of child development,[16] the child has a very good reason for internalizing the norm not to harm its parents, namely that it fears the withdrawal of their love and support. After all, the parent-child relationship is a very unequal one where in terms of power the parent is clearly favored over the child, a point which undoubtedly is not lost on the child. Needless to say, there is surely something to a crude Freudian account of why the child does not harm its parents. But need we suppose that fear of parental rejection must carry the day here? Is it not possible that the primary explanation for why the child is disposed not to harm his parents can simply be that he loves them? I believe that it is. And cases where this is so are cases where self-command grows out of love.

Now, there can be no doubt that love can give rise to self-command. Consider the case of deep friendships, a relationship which obviously involves (opaque) love. Deep friends identify with one another's good and thus take delight in one another's flourishing. Hence, whatever else they may do, they do not intentionally harm one another. They do not take advantage of one another, though it is often the case that they could easily do so without running much risk, since it is characteristic of close friends that they make themselves vulnerable to one another.[17] Of course, what is also true is that they very much desire the continuation of the friendship. That notwithstanding, what is most certainly not true is that the explanation for why friends do not harm or take advantage of one another is that they fear the loss of the friendship. This shows quite nicely that it is a mistake to think that a person's deep desire not to lose a benefit which she receives from another has to be the primary explanation for why she does not harm that individual.

It is a mistake to think this even when it comes to the child in so unequal a power relationship as the parent-child one. Love is possible among unequals in power. In particular, it is possible for a child to have the conviction that his parents will love him however he might behave. When this is the case, fear of parental rejection cannot be the primary explanation for why the child is disposed not to harm his parents. And it is reasonable to assume that the explanation here is either the child's fear of being rejected by his parents or his love for them.

In any event, when the latter is the case, what is significant is this: We have at the outset of a person's life the beginnings of one of the cornerstones of moral behavior, namely, the virtue of self-command. And it is because of the child's love for his parents that we have this. As I have remarked, morality requires that from time to time we stay our hand for the sake of others.

How we come to be disposed to act from the moral point of view is obviously tremendously important from the standpoint of moral training.

If we suppose that, as a matter of the biological make-up of human beings, morality goes against the natural bent of humans, then morality is a burden that we must bear in order to live together and, therefore, acting morally is not something that we are naturally disposed to do. But if the claims of this section are sound, then we need not make that supposition. For we have seen that at the outset of our lives displays of self-command can be seen as a deep expression of our desire not to harm others; and we have seen the role which self-command plays among friends. Thus, before acting from the moral point of view comes to be an expression of our nature as autonomous beings, it is possible for the virtue of self-command to play a significant role in our interactions and to be a deep expression of our desire to act beneficially toward selected others – our parents and friends in this case. We may think of morality as extending this virtue beyond this group of individuals to the world at large. When this virtue plays a significant role in a person's interactions with his loved ones and friends, we then have a life which is fertile for that extension. In fact this holds, *a fortiori*, if as I have argued in the preceding section, one's capacity to be weakly favorably disposed towards other persons comes in the wake of the capacity to be strongly favorably disposed toward persons.

My aim in this essay has been to show that from the standpoint of the biological make up of human beings, evolutionary biology is make up of human beings, evolutionary biology is more of an ally to morality than many have been inclined to suppose, given that we are biologically endowed with the capacity for love. I have not tried to show that however a person might be raised he will, in fact, be disposed to act morally. It is one thing to say that we have certain biologically-based capacities and dispositions; it is quite another to say that they will be actualized. Whether or not the latter is true is invariably a function of our environment. In this connection things are no different when it comes to morality.

IV

I said that I would compare the approach which I have taken in this essay with the approaches of Thomas Nagel and of Philippa Foot.

In his book *The Possibility of Altruism,*[18] Thomas Nagel endeavors to show that, *conceptually*, our taking others seriously comes in the wake of our taking ourselves seriously. Obviously enough, the aims of this essay have been far less ambitious than that. Still, our ultimate aims are similar. On Nagel's view, our taking people seriously would surely be something that we do naturally in virtue of the fact that we are rational self-reflective beings. On the view advanced here, our taking people seriously can likewise be regarded as something that we do naturally; for, in so many words, I have tried to show that morality can be grafted upon the natural affection of parental love. If Nagel has made his case, our project has not been in vain. Morality, it will turn out, has a much more secure foothold in our lives than one might have initially supposed. On the other hand, even if Nagel has failed, we still have nature, as it were, to fall back on. Nature, perhaps, is not as immutable as pure reason, but it can be pretty obstinate, which brings me to Philippa Foot.

In her provoctative essay 'Morality as a System of Hypothetical Imperatives,'[19] Foot rejects the idea that we should, on the grounds that morality has no categorical necessity to it, therefore have to worry that people might stop caring about others and defect from the moral cause. If I have argued soundly, then the claims of this essay may serve to buttress her position. For we have seen that the capacity to care about others has a rather sure footing in the lives of human beings. To be sure, it has to be nurtured and the failure to do so is likely to be disasterous. But that, in the first place, there is a capacity for caring to be nurtured is hardly insignificant. For, if nothing else, it makes being moral considerably less a matter of going against the natural bent of human beings. Our biological make-up provides us with something to work with from the moral point of view.

I believe that means-end rationality is impoverished when it comes to giving us an account of human flourishing. Any theory of morality which has this conception at its base will in the end have to acknowledge, as does Rawls himself,[20] that morality cannot be recommended to some people, period. At this very point, I want to be able to say – and this, I believe, distinguishes me from Foot – that such individuals are, either from the standpoint of natural development or of moral training, defective in someway. This essay, therefore, is part of a larger project: 'Moral Character: A Study in Motivation'. For I maintain that social interaction is the thread from which the fabric of moral character is woven.

ACKNOWLEDGEMENTS

In writing this essay, I am indebted to a number of individuals: Lynne Baker, Barnard Boxill, Stephen Darwall, Bradley Goodman, William Lycan, Thomas Nagel, Gerald Postema, Philip Quinn, Irving Thalberg and David Weissbord. A year visiting at the University of Virginia and, in particular, discussions with Daniel Devereaux and John Marshall, proved to be exceedingly fruitful in my development of views presented here. Were it not for conversations with Lycan and Nagel, the last section would not have been written. Work on this essay was supported, in part, by a fellowship from the National Humanities Center (1982–83) and an off-campus grant from the University of North Carolina at Chapel Hill (1982–83). Their support is gratefully acknowledged.

University of North Carolina at Chapel Hill

NOTES

[1] Cf. E. O. Wilson, *Sociobiology: The New Synthesis* (Cambridge University Press, 1975). Wilson's work has become something of the *locus classicus* in its field. But see also Richard Dawkins, *The Selfish Gene* (Oxford University Press, New York, 1976) and Robert L. Trivers, 'The Evolution of Reciprocal Altruism', *The Quarterly Review of Biology* (1971), 35–57.

[2] See Dawkin's, *op. cit.*, chapter 6: 'Genesmanship'. Parental care and family ties, in general, are to be explained by the theory of kin selection which, in turn, is to be explained by genes trying to maximize their numbers in the gene pool. That is, the deep explanation for why these individuals exhibit altruistic traits has to do with genes maximizing their numbers in the gene pools. Trivers, *op. cit.*, invokes the same explanation to explain the phenomena of warning calls among birds, which is an ostensibly altruistic act; for the bird who gives the call benefits others while increasing the risk of being harmed, since its call increases the chances that the predator will notice it (pp. 203–206).

[3] As Peter Singer observes, *Expanding Circle: Ethics and Sociobiology* (Farrar, Straus, & Giroux, New York, 1981): "Proof that an action we have praised had a self-interested [or some other non-altruistic] motive almost always leads us to withdraw or qualify our praise" (p. 43). The opposition to evolutionary biology often stems from the fact that it appears to take the altruism out of altruism. See Trivers' 'The Evolution of Reciprocal Altruism'.

[4] Although our projects are not quite the same, the aims of this essay are very much in keeping with Allan Gibbard's seminal essay 'Human Evolution and The Sense of Justice', *Midwest Studies in Philosophy* 7 (1982), 31–46.

[5] This distinction clearly owes its inspiration to work in philosophy of language – the theory of reference in particular. In borrowing terminology, one invariably runs the risk of having people suppose that one means for the terms to be understood in precisely the way that they were meant to be understood given the purposes for which they were introduced. I do not mean for that to happen here. *How I understand these terms is stated below in the text.* There are, it would seem, enough parallels between the way in which I am using these terms and the way in which they are used in the theory of reference to warrant my borrowing them from the latter. For a recent discussion concerning

matters of reference, see Saul Kripke, *Naming and Necessity* (Harvard University Press, Cambridge, MA, 1982) and David Lewis, 'New Work for a Theory of Universals', *Australasian Journal of Philosophy* **61** (1983), 343–377.

[6] Cf. Wilson, *Sociobiology*, chapter 27.

[7] In *On the Origin of the Species* Charles Darwin writes: ". . . I use the term Struggle for Existence in a large and metaphorical sense, including dependence of one being on another, and including (which is most important) not only the life of the individual, but success in leaving progeny." (chapter 3, par. 4.)

[8] Cf. John Bowlby, *Child Care and the Growth of Love* (Penguin Books, Baltimore, MD, 1953). He writes: ". . . what is believed to be essential for mental health is that an infant and young child should experience a warm, intimate, and continuous relationship with his mother (or permanent mother-substitute – one person who steadily 'mothers' him) in which both find satisfaction and enjoyment" (p. 13). Bowlby develops this line of thought in his *Attachment* (Basic Books, New York, 1969). Bowlby's general conclusions concerning the importance of parental love are now regarded as established. See Robert A. Hinde, 'Social Development: A Biological Approach', and Michael Rutter, 'Early Sources of Security and Competence', both in Jerome Bruner and Alison Garton (eds.), *Human Growth and Development* (Oxford University Press, New York, 1978), pp. 1–32, 33–61, respectively.

[9] See Michael Rutter, *op. cit.*, p. 38.

[10] This point is nicely made by Mary Midgley in her *Beast and Man: The Roots of Human Nature* (Cornell University Press, Ithaca, NY, 1978). For instance, as Rutter observes in 'Early Sources of Security and Competence', *op. cit.*, parental love makes it possible for children to feel comfortable in venturing out on their own, the doing of which plays a central role in their development.

[11] I have developed this line of thought in two recent essays: 'Law, Morality, and Our Psychological Nature', in Michael Bradie and David Braybrooke (eds.), *Social Justice* (Bowling Green Studies in Applied Philosophy, Bowling Green, OH, 1982), 111–123 and 'Self-Respect: Theory and Practice' in Leonard Harris (ed.), *Philosophy Born of Struggle* (Kendall Hunt Publishing Company, DuBuque, Iowa, 1983), pp. 174–89.

[12] At this point, let me simply indicate my awareness of the present debate in ethics between impartialists and personalists, recent representatives of which are, respectively: Stephen L. Darwall, *Impartial Reason* (Cornell University Press, New York, 1983) and Lawrence Blum, *Friendship, Altruism, and Morality* (Routledge and Kegan Paul, New York, 1980). As I have characterized the moral point of view, it is pretty much as impartialists conceive of it, which is understandable given my aims. For the record, let me just note that I think that personalists are not without a point.

[13] Ernest Mayr, *Evolution and the Diversity of Live* (Harvard University Press, Cambridge, MA, 1976), writes: "Natural selection says no more and no less than that certain genotypes have a greater than average statistical chance to survive and reproduce under given conditions" (p. 25). Stephen J. Gould writes: "Natural Selection is a theory of *local* adaptation to changing environments. . . . Evolution by natural selection is no more than a tracking of these changing environments by differential preservation of organisms designed to live in them." See Stephen J. Gould, 'Darwin's Untimely Burial', in Elliot Sober (ed.), *Conceptual Issues in Evolutionary Biology* (The MIT Press, Cambridge, MA, 1984), pp. 31–36. In his discussion of evolutionary stable strategies, Dawkins (*The Selfish Gene*) is clearly understanding natural selection in the way stated by Mayr and

Gould. Crudely put, then, a (set of) trait(s) would be selected for if its existence would contribute to the survival of the organisms who possess them, given the environment in which they find themselves. Traits which so contribute at one point in time need not do so at another. Hence, there is nothing about natural selection which entails that a trait selected for at time *t* will, for all future times, contribute to the survival of the organisms which possess the trait. The claim being made in the text is that, because of its referentially transparent character, parental love has contributed to the survival of the human species.

[14] For a further development of these points, see my 'Morality, The Self, and Our Natural Sentiments', in K. D. Irani and Gerald E. Myers (eds.), *Emotion: Philosophical Studies* (Haven Press, New York, 1983).

[15] On the topic of parental investiment outside of the human context, see Robert L. Trivers' essay, 'Parental Investment and Sexual Selection', in T. H. Clutton-Brock and Paul H. Harvey (eds.), *Readings in Sociobiology* (W. H. Freeman and Company Limited, San Fransisco, CA, 1978), pp. 52–97.

[16] See Freud's discussion of the origins of the sense of guilt in his *Civilization and Its Discontents*, transl. by James Strachey, (W. W. Norton and Company, New York, 1961). Helpless people, of whom the child is a paradigm example, internalize the values of those upon whom they are dependent. See chapter 7, esp. pp. 70–71, 74, 78. Freud writes: "Thus, we know of two origins of the sense of guilt: one arising from fear of an authority, and the other, later on, arising from fear of the super-ego" (p. 74).

[17] This point is developed more fully in my 'Ethical Egoism and Psychological Dispositions', *American Philosophical Quarterly* **17** (1980), 73–78.

[18] Oxford University Press, New York, NY, 1970.

[19] Collected in her *Virtues and Vices* (University of California Press, Los Angeles, 1978).

[20] Thus, in *A Theory of Justice* (Harvard University Press, Cambridge, MA) we find John Rawls making the following concession: Given their aims and wants and the peculiarities of their nature, the thin account of the good does not define reasons sufficient for them to maintain this regulative sentiment [of justice]. It has been argued that to these individuals one cannot truthfully recommend justice as a virtue. And this is surely correct, assuming such a recommendation to imply that rational grounds (identified by the thin theory) counsel this course for them as individuals' (p. 575). I think that Rawls specifically has in mind the egoist. I have attempted to blunt the force of the egoist position in my 'Ethical Egoism and Psychological Dispositions'.

PART II

EPISTEMOLOGICAL REFLECTIONS

ROBERT C. RICHARDSON

BIOLOGICAL REDUCTIONISM AND GENIC SELECTIONISM

ABSTRACT. Issues concerning the scope and viability of reductionistic explanations in biology have come recently to be discussed under the rubric of controversies over the "units" or "levels" of selection; that is, over the level or levels at which natural selection is properly said to operate. As critics have pointed out, some of the more *a priori* defenses of *genic selectionism*, the view that the sole unit of selection is the gene, have failed to demonstrate that genic selectionism provides an adequate explanatory theory. These same critics have concluded that the context dependence of genic effects due to linkage and interaction, and the resulting misrepresentation of underlying causal processes, are the cause of this deficiency. This is incorrect. Advocates of genic selectionism have not denied interaction at any variety of levels. Furthermore, to press that context dependence is sufficient grounds to impugn the explanatory adequacy of genic selectionism is untenable. Recent biological work on the coevolution of traits and historical controversies over the importance of "linkage effects" on the expression of genes suffice to press the point that context dependence is consistent with retaining lower levels of organization. Controversies over the units or levels of selection will be resolved only with due attention to the details of biological models which purport to explain evolution in terms of forces acting on lower level units and to the substantial methodological problems attendant to applying these models to actual biological systems.

Reductionism has come to hard times. Philosophical critics have repeatedly argued, not only in the case of biology, but also in the case of psychology, that reduction has little role to play in a workable scientific methodology and can only serve to distort the issues, methods, and content of the special sciences. Nonetheless, partisans remain quite unregenerate, hoping and contending that in the long run their parsimonious vision of the universe will indeed be vindicated. It surely must be admitted that there is considerable disparity between reduction as it has been described and defended by more traditional philosophical advocates and reduction insofar as it plays a role in scientific methodology and scientific theorizing. A similar disparity between reductionism as a scientific programme and reductionism as envisioned by its philosophical critics would hardly be surprising. If the gap between philosophical aspiration and scientific practice is as great as it appears to be, then the task of assessing the viability and content of reductionistic programmes has only begun.

James H. Fetzer (ed.), Sociobiology and Epistemology, 133–160.

1. BIOLOGICAL REDUCTIONISM

Within the biological sciences, the most recent forum for the discussion of reduction and reductionism has centered on the relative importance of genic, individual, and group selection in the evolution of social traits. At center stage have been a number of influential biologists, pressing that adaptations in humans no less than in other organisms can be explained as individual adaptations rather than group adaptations, and have been shaped by selection acting at the level of the gene. In his influential *Sociobiology: The New Synthesis*, E. O. Wilson wrote:

> In a Darwinist sense the organism does not live for itself. Its primary function is not even to reproduce other organisms; it reproduces genes, and it serves as their temporary carrier. . . . Natural selection is the process whereby certain genes gain representation in the following generations superior to that of other genes located at the same chromosome positions. . . . [The] individual organism is only their vehicle, part of an elaborate device to preserve and spread them with the least possible biochemical perturbation (1975, p. 3).

Though there is little doubt that selection forces generally act directly only on the organism, and therefore that the perpetuation of a gene is dependent upon the perpetuation of organism it inhabits, the claim of the genic selectionist is that the course of evolution is best charted and explained in terms of the relative (expected) fitness of alternative genes. Genes are the units of evolutionary change.

The grounds for maintaining genic selectionism to the exclusion of accounts which appeal to individuals, phenotypes, or groups are varied. They differ both in the degree to which they depend on the details of biological models and in the extent to which they divorce themselves from empirical detail. These two degrees of variation are but mirror images, and it is little surprise that philosophical attention on the issues has been concentrated at the more abstract end of the spectrum.

The most general, and most *a priori*, argument for genic selectionism is an argument from *parsimony*. There are genes whose presence is best explained in terms of selection operating at the level of the gene rather than above. The most obvious cases are among those which Richard Dawkins terms "genetic outlaws", genes which promote their own survival reasonably directly while diminishing the survival of at least some other genes elsewhere in the genome (Dawkins, 1982, chapters 8 and 9). The clearest example is that of segregation distorters. These genes influence their own proportional representation in sexual gametes, and thereby increase their proportional representation in the next generation. A gene inducing such a distortion will have a significant advantage over alternatives, just as would an individual who managed to double fecundity, even in the absence of further advantageous phenotypic effects; indeed, in the absence of serious negative effects, such a gene would be expected to spread in the population with a truly astonishing speed.[1] A second class of "outlaw" is what has come to be known as "selfish DNA" (cf. Doolittle and Sapienza, 1980, and Orgel and Crick, 1980). There is a large portion of the eukaryotic genome which is both highly repetitive and without any clear function. In fact, there are a number of repeated DNA families, and though a number of putative functions have been proposed (cf. Lewin, 1982), in but a single case has an adaptive advantage been established. Much of this repetitive DNA is likely to remain without a function, and its presence in the genome will likely be explained as simply a consequence of its being a replicator. There seems little alternative save thinking of selection for it as lying at the level of the gene. Given the need for selection acting at the level of the gene, there are principled and practical appeals for parsimony. The principled appeal favors simplicity in the abstract. It is preferable to employ as little explanatory machinery as necessary and correspondingly to acknowledge as few levels of selection as possible. G. C. Williams acknowledges that many biologists have postulated adaptations above the level of the individual (e.g., group adaptations or even adaptations at the level of the species), but contends that in "explaining adaptation, one should assume the adequacy of the simplest form of natural selection, that of alternative alleles in Mendelian populations, unless the evidence clearly shows that this theory does not suffice" (1966, p. 5). The practical appeal is one that rests on the advantages of analytical uniformity: a theory involving multiple levels of selection requires a way of comparing the relative strengths of forces operating at the respective levels (cf. Rosenberg, 1978 and 1983). Either sort of appeal to parsimony, however, is derivative and defeasible. The force of each depends upon the demonstrable adequacy of lower level theories, and presupposes that evolutionary change can be adequately explained in all cases in terms of selection operating to optimize the (expected) fitness of genetic units. It is the latter which is at issue between genic selectionists and their opponents, and it is hard to imagine that controversies over the 'units of selection' would survive its resolution or that any further appeal parsimony would then be anything but otiose.

An equally general argument, and only slighly less *a priori* in character, rests on the possibility of reconstruing any dynamic model using higher level units in terms of a model which assigns fitness values only to genic units. In a passage often cited by critics,[2] G. C. Williams offers the following defense of one-locus models of natural selection.

The unity of the genotype and the functional subordination of the individual genes to each other and to their surroundings would seem, at first sight, to invalidate the one-locus model of natural selection. Actually these considerations do not bear on the basic postulates of the theory. No matter how functionally dependent a gene may be, and no matter how complicated its interactions with other genes and environmental factors, it must always be true that a given gene substitution will have an arithmentic mean effect on fitness in any population. One allele can always be regarded as having a certain selection coefficient relative to another at the same locus at any given point in time (1966, p. 57).

The point Williams is making is captured in a metaphor used by Richard Dawkins (1976). Dawkins acknowledges the "whole set of genes in a body constitutes a kind of genetic climate or background, modifying and influencing the effects of any particular gene." He likens the set of genes constitutive of the genotype to a rowing team. Given any seven team mates, one can readily find the optimal team by trying out the remaining candidates in the open slot. By varying all the positions, one could eventually fix on the best team. An individual member would be singled out as a good oarsman if, on the average, he tended to be in winning boats. "A gene which cooperates well with ... the genes in the whole rest of the gene pool", Dawkins concludes, "will tend to have an advantage" (1976, pp. 40–42). The relative fitness of any gene relative to other alleles is thus partly a function of the other genes in the gene pool and their relative frequency, as well as factors such as chromosomal inversions common in the population which affect recombination. But, given such complications, Williams and Dawkins nonetheless assure us there will be a determinate difference in fitness for any set of alternative alleles within the population at any given time. "Adaptation can thus," Williams concludes, "be attributed to the effect of selection acting independently at each locus" (1966, p. 57). It is then only a small step from Williams' reminder that "group selection and biotic adaptation are more onerous principles than genic selection and organic adaptation ... [which] should only be invoked when the simpler explanation is clearly inadequate" (1966, p. 124) – that is, from an appeal to parsimony – to the ultimate point Williams wishes to enforce: "A gene is selected on one basis only, its average effectiveness in producing individuals able to maximize the gene's representation in future generations" (Williams 1966, p. 251).

2. INTERACTION AND EXPLANATORY SUFFICIENCY

The supposition which evidently underlies arguments of this last sort is that if evolutionary changes can be described and tracked in terms of changes in gene frequency, then there must be an adequate dynamic account explaining those changes in terms of forces – captured quantitatively as selection coefficients – acting on genic units.

Similar suppositions can be found in the philosophical literature. Kenneth Schaffner, e.g., has elaborated a model for reduction designed to fit the biological sciences. He claims that one element in successful reductions is the construction of "reduction functions" which serve as "dictionary entries":

> (1) From the context of the theory to be reduced one must be able to take any of the terms . . . and univocally replace these by terms or combinations of terms drawn from the reducing theory. (2) From the context of the reducing theory one must be able to pick out combinations of terms which will unambiguously yield equivalents of those terms in the reduced theory which do not appear in the reducing theory (Schaffner, 1974, p. 622).

For, say, an expression referring to a gene for a specific coat color in guinea pigs, we determine which entities lie within its extension, and then simply construct a chemical description of what we find, no matter how heterogeneous and disjoint that may prove to be. Failing to find coreferring expressions is out of the question, though they might be unwieldly in the extreme. Schaffner is aware of this possibility and consequently adds quickly that his approach is "only an ideal standard for accomplished reductions, and does not characterize the research programmes of molecular biologists" (Schaffner, 1974, p. 613). Nonetheless, appealing to the possibility of constructing such "reduction functions", he maintains that "a reduced theory is *in principle* eliminable" (Schaffner, 1974, p. 615).[3]

The plausibility of the proposal evaporates once we ask for what purposes the "reduced theory" is "eliminable". The prospect of constructing such "reduction functions" surely warrants the claim that we could (if only 'in principle') *redescribe* the mechanisms of Mendelian inheritance in molecular terms. But we have gained little and lost much. The prospect of such redescription does nothing even to suggest that the laws so redescribed remain lawlike. We may have an acceptable description of the processes, but lack altogether the predictive and explanatory resources which it is our goal to acquire. Moreover, even if the laws so reformed remain lawlike, there is little to suggest that, so redescribed, the descriptions deployed will be sufficiently simple and uniform for that law to count as a law within the lower level theory. Even if we were to have a law, it would lack the sort of systematic integration which alone would make it useful.

Critics of Williams have elegantly isolated a parallel flaw in his argument for the sufficiency of genic selectionism. In a pair of seminal papers, William Wimsatt has criticized this reductionistic vision as being theoretically unwarranted and methodologically untenable (cf. 1980a and 1980b). He argues that Williams can promise us little more than "genetic bookkeeping" in which the complexities due to interactions at higher levels are redescribed, and that genic selectionism cannot provide a predictive (or explanatory) theory at all. A map of gene frequencies across time may be an acceptable *description* of population changes, and it may even be intended as a basis for trend extrapolation, but it cannot give us an explanation of the relevant changes and will not serve as a reliable predictive guide. To enforce these conclusions – and in particular, to support the view that as a predictive variable, genic frequencies are inadequate – Wimsatt asks us to consider a case described by R. C. Lewontin (1974), based on studies by Lewontin and M. J. White (1960) of a field population of the grasshopper *Moraba scurra*. Figure 1 is a graph

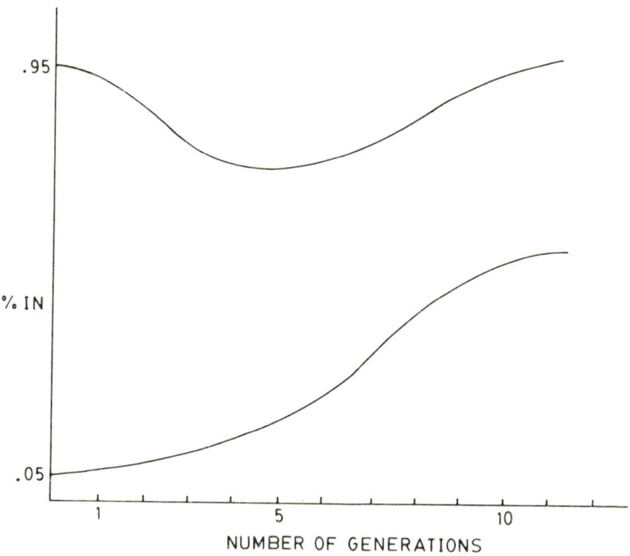

Fig. 1. The frequency of inversion polymorphisms in populations of the grasshopper *Moraba scurra* based on work by Lewontin and White (1960). Graphs represent the average behavior of replicates. Adapted from Lewontin (1974).

of changes in gene frequencies of two different allelic pairs across generations from two initial points (0.05 and 0.95).[4] In the lower curve, single locus allelic frequencies are adequate to predict subsequent frequencies (at least for the limited range presented). We have a simple increasing function dependent on prior frequency. In contrast, consider what happens to the frequencies of the alleles charted in the upper curve. That case is quite unlike the allelic pair charted in the lower curve. The value of the gene frequency is insufficient to predict changes in gene frequency; for, given any value between the initial point and the minimum reached in the fourth generation, while that value is followed by a decrease prior to the fourth generation (e.g., at generation two), it is followed by an increase subsequent to that generation (e.g., generation eight). A prediction of subsequent frequency cannot therefore be made on the basis of gene frequencies alone. Allelic frequencies are not even adequate to predict the direction of change – whether the relevant alleles are going to increase or decrease – much less the magnitude of change. Lewontin concludes his own discussion of the case:

> . . . the gene frequency alone is not a *sufficient dimension*, because we cannot describe the evolution of the system unambiguously in terms of that dimension and a set of parameters whose values are determined externally to the system. We are in the same situation as the pilot of a space capsule who knows only his position and velocity. He cannot plot his course unless he also knows whether he is speeding up or slowing down. Some dimensions are missing from our description of the evolution of the inversion system, without which our theory is dynamically insufficient (Lewontin, 1984, p. 275).

The direction of change is, however, predictable as a function of two alleles at two loci. The two curves actually describe interacting loci: Whether the changes in the upper curve are positive or negative is partially determined by the frequencies charted in the lower curve. Prior to the fourth generation, the frequency of this second allele is too low to effect an increase in the frequency of the allele charted on the upper curve. After that point, the allele is favored because of the increased frequency of the second allele. In short: though the single parameter, gene frequency at one locus, yields no determinate function mapping gene frequency at one time to gene frequency at a subsequent time, there is such a function using two variables specifying the frequencies at two loci. Lewontin concludes: "The indeterminacy and the self-contradictory paths of . . . change in these populations are reflections of the dimensional insufficiency of a single-locus treatment" (1974, p. 280). Similarly, Wimsatt concludes that Williams' proposal fails to yield a "deterministic theory" of evolutionary change "because epistatic interactions among loci prevent local estimates of fitness at single loci from being

developmental process] is encapsulated completely in the original instructions [encoded in the genes] and awaits only a mechanical unfolding" (1983, p. 18). That Dawkins grants such interaction is clear. That he thinks it still possible to maintain the gene as the unit of selection is also clear. What is less obviously correct, once such developmental interaction is granted, is whether it will still be open to us to think of the gene as the unit of selection.

This serves as a wedge that threatens the critique of Wimsatt's as well as that of Sober's and Lewontin's. In the case which Wimsatt relies on, Williams could freely grant the insufficiency of gene frequencies as a predictive measure. Williams clearly contends that "the population gene pool constitutes one aspect of the environment in which the natural selection of alternative alleles must take place" (1966, p. 61). He points out that this is what is implied in discussions of the integration of the genotype and the coadaptation of genes, claiming that "recognition of this principle is perfectly consistent with the theory that natural selection of alternative alleles is the only force responsible for the production and maintenance of adaptation" (*loc. cit.*). On Williams' view, alleles have determinate selection coefficients only within a specific genetic environment. Given this, he is antecedently committed to denying that one could predict selection coefficients from frequency measures at single loci; indeed, were it so, his recognition of "environmental levels" would be wholly incoherent. To suppose that changes in gene frequency should be a function of prior gene frequency alone must rest on an assumption that the relative fitness of the relevant genes is either constant or a function of the frequency of genes at that locus. Williams is no more committed to this assumption than is Wimsatt. Yet, without it, the appeal to the case of *Moraba scurra* is hardly probative. Sober's and Lewontin's attack on Williams is not obviously on a better footing. In the case they use to support the principle of genetic relativity, that of heterozygote superiority, the key flaw in the genic selectionism they criticize is ignoring the 'genic environment' and averaging fitness values for a gene (type). It will naturally turn out that any sensitivity to interaction effects is obscured. It would equally well be open to the genic selectionist to insist that we must acknowledge the differential 'fitness' assignments in different genic contexts. There would, of course, be a tremendous cost in terms of computational complexity, and for that reason no such calculations are generally adopted. The general point remains, however: the analytical methods deployed in developing the argument ignores the commitment to the relativity of fitness values.

None of this shows that Williams is right, much less that a commitment to genic selectionism is defensible. Neither does it show that genic selectionism is capable of presenting a workable explanatory account of the workings of natural selection. The conclusion warranted is more modest. If the appeal to context sensitivity in the form of interaction effects, whether between alleles at the same locus or between loci, is consistent with genic selectionism, then there is a response available to the criticisms of Wimsatt on the one hand, and of Sober and Lewontin on the other. What critics have shown is, first, that the mere possibility of genic redescription of evolutionary changes does not show that the sole mechanism is selection at the genic level[5], and, second, that any theory which fails to accommodate interaction effects will be dynamically insufficient and theoretically untenable. The question is then whether gene-level interaction effects preclude genic selectionism as an explanatory theory. More generally, the issue is whether interaction between entities at a level which has the effect of modifying fitness values for those entities is sufficient to insure that selection effects at that level tell at most an incomplete story. Less precisely but more concisely, does interaction suffice to require higher levels of selection?

3. BIOLOGICAL REDUCTIONISM AND INTERACTION

Genic interaction is a case of simple context-dependence of effects. What genic interaction amounts to, after all, is the sensitivity of the effects of a gene on the phenotype (and specifically on the fitness of the phenotype) to the presence of other genes at the same or other loci. Critics can then be seen to be urging that the context-dependence of genic effects is indicative of the explanatory inadequacy of genic selectionism, and, by implication, that genic selectionism would be adequate only if genic effects were insensitive to context.

This general commitment is most in evidence in Wimsatt's definition of a unit of selection. He writes:

> A *unit of selection* is any entity for which there is heritable *context-independent* variance in fitness among entities at the level which does not appear as heritable context-independent variance in fitness (and thus, for which the variance in fitness is *context-dependent*) at any lower level or organization (Wimsatt, 1980a, p. 236; cf. Wimsatt, 1980b, p. 144).

I do not propose to get deeply involved in the more technical issues which the definition invites us to consider.[6] There are essentially three distinctive conditions. Each is intended to be necessary. Jointly they are to be sufficient.

First, then, in order for there to be evolution occurring under the influence (cf. Lewontin, 1970). In the shorthand form, there must be heritable variation in fitness. This tells us virtually nothing about the level at which selection is occurring, or the mode of transmission accounting for the heritability of variation. The remaining two components fill this gap.

Second, then, in order for there to be evolution occurring under the influence of natural selection at a given level of organization – briefly, for the unit of selection to lie at the level – variations in fitness at that level must not be analyzable exhaustively as the aggregate effect of variations in fitness of entities at any lower level.[7] When, e.g., an explanation is offered for warning cries in birds in terms of mutual benefit, or when the classic limitations on clutch size, are explained as optimal for the individual, then the explanation in terms of individual advantage is taken to screen off and render impotent any explanation in terms of higher level units – say, in terms of benefit to the flock or community (cf. Trivers, 1971). The effect of the condition is to guard against higher level artifacts (see Sober, 1980, *R*3, for a clear discussion of this condition). Thus, if there is a component of the heritable variations in fitness that is not an aggregate effect of heritable variations in fitness at lower levels, then there is selection occurring at that level or higher.

The third condition, context-independence, provides a condition under which there will likewise be *no* selection at higher levels. If all the heritable variation in fitness at a given level that is not the aggregative effect of these variations at lower levels is insensitive to context, then, at all higher levels, any heritable variation in fitness will be an aggregative effect of variations at that level or below. "If variance in fitness is totally additive at a given level of organization over a given range of conditions on the environment and the system, then, under those conditions there are no higher-level units of selection!" (Wimsatt, 1980a, pp. 236–237).[8]

The second and third conditions are mirror images, using the same analytical machinery to determine, respectively, whether lower level selection is sufficient and whether higher level selection is unnecessary. The third condition is more central for our interests, since it stipulates that a given level of selection is insufficient to explain the evolution occurring due to natural selection if there is context-dependence at that level.[9] As Wimsatt remarks, context-dependence is not sufficient to guarantee that there are higher level units of selection (Wimsatt, 1980b, p. 147). There is a further condition to be met; *viz.*, that the remaining variations in fitness must be context-independent at some higher level. He thus does not hold that any context dependent effects must be context independent at some level of organization. What would we do with context dependent effects that are not context independent at any higher level? Wimsatt is apparently committed to saying that any residue is simply not due to natural selection operating. There is a certain degree of tension in such a position. But let us set this prospect aside. We can restrict our attention to heritable variation in fitness that is factorable into contributions which are, at some level or other, context independent and therefore which is totally context-independent at the highest level. It then will follow that the presence of context-*dependent* effects at a level will be sufficient to guarantee that there are higher levels of selection.

Sober apparently rejects the condition of context-independence, as when he remarks, apparently in criticizing Wimsatt:

> What more is there to group selection beyond there being heritable variation in the fitness of groups? One natural suggestion is this: if the change in gene frequency one gets would not have happened if one had a single panmictic population, then one has group selection. . . . The idea here . . . is that group selection exists whenever *population structure* affects reproductive success. . . . [It] seems clear that the proposed definition is extremely liberal, in that it makes group selection a commonplace in evolution. It is a virtually universal fact about fitness that an organism's fitness depends on what the other organisms are like in the population. . . . group selection must involve more than the fact that fitness values are *context sensitive* (Sober, 1980, pp. 102–103).

Similarly, Sober and Lewontin remark:

> If the fitness of *X* depends on genetic context, this may mean that there is a single selection process at some higher level, *or* it may mean that there are several different selection processes at the level of *X*. Context sensitivity does not suffice for there to be selection at a higher level (1982, p. 170).

They suggest that some further condition is needed, suggesting: "What seems to be required, roughly, is that some common causal influence impinge on the organisms" (1982, p. 171). The example they deploy is some help. They ask us to consider a gene which is lethal when it occurs alone, but has no effect when in the presence of a suppressor gene. If there are then two subpopulations, one in which the suppressor is unviersal and one in which it is wholly absent, and therefore one in which the lethal is unexpressed and one in which it is, then, they contend, we have two different selection forces operating and no higher level units of selection.[10] But their differences with Wimsatt are far fewer than their points of agreement. They are after all committed to the following principle: if there is context dependence at a given level, and there is but a single selection force operating at that level, then there must be units of selection at some higher level. It may be that the possibility of there being context-dependent effects which are not context-independent at some higher level is, in Wimsatt's hands, merely an acknowledgment of the possibility of multiple causal influences. If so, then there is no divergence of view at all. But in any case, their differences are not as great as Sober and Lewontin suggest.[11]

The upshot of this rather tangled nest of issues is this. The existence of context-dependent effects modifying the fitness of an entity *X*, on Wimsatt's view, is sufficient to show that an explanation of evolutionary changes in terms of selection on *X* (or below that level) will fail to satisfy the condition of dynamic sufficiency. There will therefore either be *no* exlanation for some part of the evolutionary changes or an explanation only at higher levels of selection. The existence of such context-dependent effects, on Sober's and Lewontin's view, has the same consequence provided that there is a single and uniform selection force operating on all the entities in the domain.

Neither view has a great deal to recommend it. This can be seen by turning to two different cases, each of which suggests that we should resist this common emphasis on context-dependence and interaction.

(i) *Coevolution and Frequency Dependence.* In *On the Origin of Species* (1859), Charles Darwin began with the observation that any view on the origin of species "would be unsatisfactory, until it could be shown how the innumerable species inhabiting this world have been modified, so as to acquire that perfection of structure and coadaptation which most justly excites our admiration" (p. 3). His own solution, of course, was the mechanism of natural selection. It was integral to his own defense of natural selection (though not of evolution) that the "coadaptations of organic beings to each other and to their physical conditions of life" should be developed by a gradual process of "modification and coadaptation" (p. 4). It was equally integral to his views that natural selection was generally a force acting on individuals rather than groups, communities, or species.[12] Even at the birth of evolutionary biology, therefore, a commitment to acknowledging interaction and interdependence was regarded as compatible with – indeed requiring – recourse only to lower levels of organization and selection only at those levels.

More recent work in evolutionary biology and evolutionary ecology has not changed the fundamental picture, though it has given us a more systematic treatment and analysis of the dynamics of the situation.[13] The fundamental issues concern the interaction of distinct populations, and distinct species, and how this interaction can and does shape the participants. The development of the complex interdependencies of the sort that is ubiquitous in ecological systems cannot be understood without attention to the topic of coevolution. I propose to concentrate on the case of *purely density-dependent coevolution.* This is by no means the most general case. To focus on this model means assuming that, within populations, the traits manifested are constant (e.g., that behavior patterns and coloration are uniform within the populations) and, thus, that the only relevant variable affecting the dynamics of the system is population sizes.[14] A more complete treatment would demand specification of both variables. Given our limited goals, the added complexity would not pay off, and, in any case, would not alter our fundamental conclusions.

Specifically, we shall concentrate on a four-species system, with limited resources. An example of the sort of system which will be dealt with would be one with two species of hummingbird, each specialized for their respective flowers, with the plants in direct competition (e.g., for space or resources) so that they tend to exclude one another. The interaction between each hummingbird species and the plant it feeds on can be regarded as a case of predator-prey coevolution, with negative density dependence; that is, an increase in the population size for the plant will have a positive effect on the size of the bird population. The interaction between the two plant species is to be regarded as a case of density dependent co-evolution; that is, it is such that an increase in the size or frequency of one species within the range will have a negative effect on the size or frequency of the second species. The two bird populations are not in competition for resources at all.

The system may be depicted as in Figure 2, following Richard Levins (1974). A line ending with an arrow indicates a positive effect, so that, e.g., the effect of **a** on **A** is such that an increase in the number or frequency of **a** – that is, an increase in $\mathbf{N_a}$ or in $\mathbf{p_a}$ [$= \mathbf{N_a}/(\mathbf{N_a} + \mathbf{N_b})$] – will yield a positive effect on the number or frequency of **A**. A line ending with a dot indicates a negative effect, so that, e.g., an increase in $\mathbf{N_a}$ will have a negative effect on $\mathbf{N_b}$ and, accordingly, as $\mathbf{p_a}$ approaches unity, $\mathbf{p_b}$ approaches zero.

It would be possible to attempt to analyze this system as a two-way competition between **Aa** and **Bb**, since it is possible independently to specify equilibrium values within each subcommunity.[15] Such a proposal, however, misrepresents the situation. The direct competition within the system is between **a** and **b**. The only effect of **A** on **B** or of **B** on **A** is

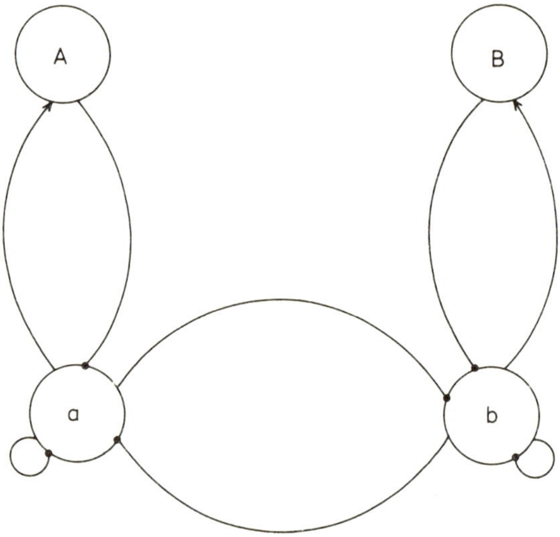

Fig. 2. A diagrammatic representation following Levins (1974) of interactions in a four species system. A line of influence ending with an arrow indicates a positive effect, and a line ending with a dot indicates a negative effect. E.g., **a** has a positive influence on **A** in that an increase in the size (N_a) or proportion (p_a) of the **a** population will cause a corresponding increase in **A**; **a** has a negative influence on **b** in that an increase in **a** will cause a decrease in **b**. The interaction between **a** and **A** is one of predator-prey coevolution, as is that between **b** and **B**. The interaction between **a** and **b** is one of frequency dependent coevolution.

indirect, mediated by their moderating influence on the prey populations. Insofar as the degree of influence of **A** on **a** and of **B** on **b** is minimal in comparison to other determinants of N_a and N_b[16], the dynamics of the system will be dominated by these other determinants. That is to say, the dynamics of the system as a whole will be determined by such factors as the relative efficiency of **a** and **b** in resource utilization and productivity. Apparent competition between **A** and **B** will be merely epiphenomenal.

What happens as we introduce perturbations into the system? These can be read directly off the diagram. Suppose, e.g., that N_a and therefore p_a increases. This will have a positive influence on N_A and an inhibiting influence on N_b, which will in turn effect a decrease in N_B. The net influence on **A** and **B** will be an increase in p_A relative to p_B. A similar effect would result from an increase in N_b.

Suppose now that we have a system which, at equilibrium, reflects the dynamics of the model. Suppose further that, from whatever cause, $\mathbf{p_a}$ is increasing and $\mathbf{p_b}$ is correspondingly decreasing, and that $\mathbf{p_a}$ is initially low in comparison to $\mathbf{p_b}$. Suppose, finally, that $\mathbf{N_A}$ and $\mathbf{N_B}$ are significantly below their optimum values for the relevant values of $\mathbf{N_a}$ and $\mathbf{N_b}$, and therefore that the system is *not* in equilibrium. What will happen as time goes on? Well, **a** will slowly displace **b** in the population, at least up to a limit. Initially, both $\mathbf{N_A}$ and $\mathbf{N_B}$ will grow, but since $\mathbf{p_a} \ll \mathbf{p_b}$, it is reasonable to assume, other things being equal, that the rate of increase of $\mathbf{N_B}$ will be higher than that of $\mathbf{N_A}$. Accordingly, the value of $\mathbf{p_A}$ will decrease and that for $\mathbf{p_B}$ will increase. The approach to equilibrium values for $\mathbf{N_A}$ and $\mathbf{N_B}$ will be relatively fast, with the result that $\mathbf{p_A}$ and $\mathbf{p_B}$ will temporarilly stabilize. The stability is, however, short-lived. Since the value of $\mathbf{p_a}$ relative to $\mathbf{p_b}$ is increasing (though more slowly than had been the relative values of $\mathbf{p_A}$ and $\mathbf{p_B}$), the trend will reverse. The value for $\mathbf{p_A}$ will increase and that for $\mathbf{p_B}$ will decrease, following the rise of $\mathbf{p_a}$ relative to $\mathbf{p_b}$. This will continue until the equilibrium values for the system are reached.

Only a little examination will suffice to convince the reader that the graph of the population dynamics just described is that given in Figure 1 above. The interactions within this model ecological system are analogues of the interactions within the genic system described by Lewontin and used by Wimsatt. Wimsatt uses the argument to show that gene frequencies are not dynamically sufficient and concludes that we must move to an examination at higher levels of organization in order to account for the evolutionary properties of the system. Interaction of this sort, he contends, requires an acknowledgment of higher levels of selection. Yet the interactions within this model ecological system suffice to show that the values for $\mathbf{p_A}$ and $\mathbf{p_B}$ are also dynamically insufficient. Yet it would be wrong, as we have seen, to conclude on this ground alone that we should shift to a higher level in order to understand the dynamics of this system. The dynamics of the system are properly understood as a consequence of a four-species interaction (i.e., between **A**, **B**, **a**, and **b**) and *not* as an interaction between two communities (i.e., **Aa** and **Bb**).

The conclusion that is warranted appears to be quite general. The existence of complex interactions, of a sort that insures context-dependent effects, does not suffice to drive us to acknowledge higher level units of selection. Since the example reflects a single and uniform selection force along each direction of exchange, the conclusion is as much at odds with the proposal of Sober and Lewontin as it is with that of Wimsatt. A second case will reinforce the point.

(ii) *The Position Effect and Unit Characters.* In 1913, William Bateson claimed that "the essence of Mendelism lies in the discovery of the existence of unit characters of factors" (1913, p. 3). The notion of a unit character developed from an amalgam of two crucial Mendelian concepts. Bateson viewed the result of Mendel's work as requiring that there be transmissable units, or "germ cells", which were unaffected by the process of sexual reproduction, and that, when there are alternative "allelomorphs", there should be "antagonistic characters" at most one of which is expressed. The *unit character*, accordingly, combined the idea that there is a transmissable *unit* with the idea that these units are expressed in unique *characters* (or not expressed at all). One could hardly ask for a clearer and more definite commitment to the view that the *organization* of the genome is irrelevant. Indeed, even in the face of the discovery of non-Mendelian ratios, Bateson (together with Punnett) denied the effects of linkage (i.e., chromosomal organization), contending that the deviant ratios could be explained as the result of differential multiplication of the relevant cells in pre-meiotic divisions. The dispute was, of course, finally resolved in favor of Morgan and his followers. Organization was taken to affect the ratios in the transmission of traits.

The notion of a unit character survived in another form, however; for even if chromosomal organization was seen to affect transmission, the role or organization in the *expression* of genes went unrecognized for some years.[17] Yet by 1932, Muller and Prokofyeva were able to assert:

> [The] genes are, from the point of view of their functions in determining the characters of the organism, not discontinuous, in that neighboring genes enter into special relations with one another, with the resultant formation of a gene-system. . . . The basis of genetic determination of the characters of an organism cannot be stated completely by a mere listing of the individual genes which the organism contains; the arrangement is itself, in effect, genic in nature (cited in Carlson, 1966, pp. 149–150).

What happened in the intervening years was the discovery and establishment of what is known as the "position effect". Until the critical work of A. H. Sturtevant and T. H. Morgan in the 1920's, position on the chromosome was thought to have little or no effect on the expression of genes. Sturtevant was intent on studying gene mutation by studying a strain of *Drosophila* with abnormal eye structure. The 'mutation' was a relatively frequent one that had been identified as a sex-linked dominant[18] having the effect of diminishing the number of facets in the eye, and therefore the overall size as well. It had been known for some time that, in spite of its dominance, there was occasional reversion to the 'wild type' with a round eye, and this was taken to be a reverse mutation. It had also been found that homozygous bar was capable of giving rise to a more deviant form, initially dubbed "ultra-bar" in which the number of facets was even further reduced. Ultra-bar, in turn, occasionally reverted to both the 'wild type' and to a bar type. By studying reversion frequencies and the results of crossing-over, Sturtevant showed that the reversions were not reverse mutations at all, but, rather, the results of unequal crossing-over at the bar locus. As a result of an unequal exchange of chromosomal material, material originally lying on distinct but homologous chromosomes came to lie in adjacent positions on a single chromosome. Later microscopic studies (see Bridges, 1936) of the banding patterns in salivary chromsomes showed that a homozygous bar individual had a sequence in each chromosome represented as

15 · 16A · 16A · 16B.

Unequal crossing-over of a homozygous bar, in turn, could yield one chromosome with a configuration involving a triple **16A** segment

15 · 16A · 16A · 16A · 16B,

and a second with a 'wild type' configuration,

15 · 16A · 16B

The presence of a triple **16A** sequence was responsible for 'ultra bar', which Sturtevant renamed "double bar".[19] Unequal crossing over thus was capable of explaining both the tendency for bar individuals to give rise occasionally to the more extreme double bar and the reversion to 'wild type'.

In the process of investigating this phenomenon, Sturtevant and Morgan took note of an even more striking and surprising phenomenon. Comparison of the number of facets showed that the heterozygote with a double bar and 'wild type' chromosome had significantly fewer facets than a homozygous bar individual. In the scheme of representation above, each type of individual has a total of four **16A** segments.[20] Sturtevant concluded that there was a "position effect": "two bar allelomorphs lying in the same chromosome are more effective than are the same two allelomorphs when they lie in opposite chromosomes" (Sturtevant, 1925, p. 140).

Dobzhansky (1932) later showed that the original bar phenotype involves a position effect as well. The bar 'mutation' is not a mutation at all, but a result of unequal crossing-over, and the fact that the dominance in the heterozygote is incomplete is a position effect. The lack of proximity of the number **15** segment to one of the **16A** segments releases the latter. Hence, the bar phenotype is not merely a matter of increased dosage due to a gene duplication, but a consequence of the interaction between genes at adjacent loci. The point is not merely that there are polygenic influences – that had been recognized for some time – but that the relative location was itself a significant factor in determining the expression of genes.

The discovery of the position effect did indeed institute a significant reshaping of the classical conception of the gene. Few, however were willing to go as far as was Richard Goldschmidt, who appealed to the position effect in order to justify a wholesale rejection of the classical conception. He wrote in 1937:

> . . . the time has come to acknowledge that gene mutations have as little existence as genes themselves. . . . The idea of a position effect, made to save the gene concept, will also have to disappear when it is recognized that the position effect is actually identical with what was called a gene. The chromosome as a unit will be found to control normal development or wild type. The changes of the correct order within its chain produce deviations from normal development, called mutants. Though they are localized, there is no such thing as a gene and certainly no wild type allelomorph (cited in Carlson, 1966, p. 125).

Goldschmidt urged the abandonment of the notion of the gene and the effect it has on development in favor of a view which took the chromosome as a single developmental unit, with all 'mutations' being chromosomal rearrangements and the ensuing position effects. The proposal received little attention. Advocates of the classical model of the gene instead satisfied themselves with adhering to the gene as the fundamental unit and incorporating the position effect as yet another form of interaction between these units.

To follow Goldschmidt, I submit, would have been ill-advised, even from the perspective of theorists at the time. Their reaction was entirely appropriate. It would have been ill-advised methodologically and theoretically to abandon the gene as the fundamental unit of analysis in the face of evidence concerning the position effect. The principle that complex interactions, of a sort that insures context-dependent effects, require higher level units of selection is at base a principle that requires higher level units in the face of significant interaction. It application in the case of the position effect would have led us to side with Goldschmidt. The proper conclusion is that the principle is unsound. The acknowledgment of even very intricate forms of causal interaction does not require that we shift to a higher level. We must acknowledge the interaction, to be sure, but we are not well-advised to abandon the fundamental units in favor of a more 'macro' analysis on these grounds alone. This can be reinforced by noting that what is at issue is not merely some loose analogy. The link between transmission and selection is development. Accordingly, the acceptance of interaction in both transmission and expression, while retaining the gene as the fundamental unit, is a recognition that interaction alone does not dictate the recognition of higher level units for selection[21] either.

4. CONCLUSION: GENIC SELECTIONISM

What, then, should we conclude concerning genic selectionism? The more *a priori* arguments favoring genic selectionism are inadequate. They fail, more specifically, to show that genic selectionism gives an acceptable accounting of the relevant causal determinants, and hence fail to show that it provides an acceptable explanation of evolutionary changes. Critics, however, have hardly come out in much better shape. They have correctly seen that the case for genic selectionism provided by Williams, and to a lesser extent by Dawkins, is inadequate. Unfortunately, they have localized the supposed source of this inadequacy in a failure to recognize context-dependent effects due to interaction and linkage. We now have some reason to hold that the mere existence of context-dependence is not enough to support the demand for higher level units. It seems we are left with a standoff, and to some extent we are. What I wish to suggest in concluding is the direction we must turn to resolve the impass.

The case for genic selectionism does not rest on such *a priori* gounds alone. Biologists have also developed detailed explanatory models. These are dynamic models, generally mathematical in form, relying exclusively on lower level units, and purporting to be sufficient for explaining the evolution of social traits (for a general treatment, see Richardson (forthcoming)). Such models specify the direction and form which selection at the lower levels with take, given determinate parameter assignments to variables for such things as the degree of relatedness between interacting conspecifics, the benefit to recipients, and the cost to the agent. One source for such work involves the utilization of classical game theoretic models to predict optimal evolutionary strategies on the basis of individual advantage. The premier proponent of such an approach is surely John Maynard Smith (1982, 1983, 1984; also Dawkins, 1976; and Axelrod and Hamilton, 1981; for a critical assessment on methodological grounds, Richardson (1984)).

The most powerful work in this vein is perhaps W. D. Hamilton's landmark treatment of kin selection in the evolution of social 'altruism'. Hamilton develops an extension of classical models of selection in order to obtain a model capable of explaining the evolution of traits which are not to the advantage of individuals manifesting them. The point of the model is, then, to demonstrate that the direction of selection at lower levels is at least in qualitative agreement with information on animal behavior. The key construct is the notion of *inclusive fitness*. In a somewhat simplified form this can be regarded as composed of two elements – the personal fitness, which is the fitness of the organism independently of any social contribution, and the differential social fitness, which is the net fitness contribution of the indvidual to genetic relatives. This latter contribution in turn is a complex with two principal components: it is the individual's contribution to genetic relatives less the individual's contribution to non-relatives. Given such components, Hamilton shows that it is possible to determine whether there is a net advantage conferred by a given trait. The issue is somewhat controversial, but the proper construal of the model would appear to treat it as locating selection at a level lower than the individual. The unit of selection is *genic*, since the question is whether a trait conferring a distinctive social advantage will spread through a population.[22] Hamilton's model construes selection forces as operating to maximize the representation of particular genes or gene complexes in future generations, allowing that such forces may fail to maximize individual fitness. The importance of kinship relations is simply one of developing a mechanism which would give a quantitative estimate of genetic fitness.[23] Given such an estimate, and the appropriate parameter values, the direction and degree of selection based on inclusive fitness will be fixed. Whether these will prove to be dynamically sufficient once developed is uncertain. If they do not, then we should expect a significant role for higher level units of selection.

Evaluating the strength of arguments of this last, decidedly *a posteriori* type, is more difficult and more challenging. There are serious, deep and interesting methodological considerations needing to be addressed, not the least of which are questions over whether the models incorporate the relevant parameters and whether the parameter values assigned are realistic ones. Many of these latter issues depend on whether we can obtain reliable quantitative estimates of expected fitness[24] in order to gain applications of our explanatory models which provide significant tests of the models. These in turn invite an exploration of the role of optimality models in biological theorizing, since these provide the most promising basis for such estimates.[25] Issues such as these raise questions of scientific methodology and biological theory that are largely unsettled, and constitute a relatively untapped area for philosophical work.

ACKNOWLEDGEMENTS

The research for this paper was supported by a grant from the National Science Foundation.

I am grateful for extended discussions with Robert Brandon, Richard Burian, and William Wimsatt. Each has had a significant influence on my views on these subjects.

University of Cincinnati

NOTES

[1] R. C. Lewontin shows us that (in the absence of stochastic factors) a doubling of fecundity in one mutant individual in a population of one hundred would go nearly to fixation in a dozen generations. Without further information concerning the pressures to which the population is subject – whether, e.g., there are predators concentrating on immature forms with a density dependent strategy – there will be no determinate answer to the question whether the average adaptedness of the population will be higher or lower after the mutant has spread to fixation than before. See Lewontin (1977), pp. 10ff.

[2] It is William Wimsatt that first drew my attention to the argument, and, following his lead, it has become common to construe this argument as crucial to the case for genic selectionism. I think Wimsatt is right to reject the argument, for reasons which will be recounted briefly later, but I am less certain about the centrality of the argument to the case for genic selectionism.

[3] Schaffner is not alone in uncritically accepting such views. For other cases, see Nagel (1961), Hellman and Thompson (1975), and Causey (1977).

[4] Actually, we are charting the frequencies of inversions – constellations of genes which segregate essentially as a unit – rather than genic frequencies. This does not

matter. The behavior of inversions is such as to make them approximate Mendelian units as nearly as anything does; that is, they behave as genes.

[5] Stochastic processes – in particular genetic drift – are not at issue, and can be accommodated to any of the competitors.

[6] In particular, I want to avoid issues over the analysis of variance. I do think it necessary to note, however, that I regard Wimsatt's appeal to *variance* in the definition as a mistake, and will amend my discussion accordingly. A full examination would need to take account of Wimsatt's use of variance, and especially the justification for including context-dependence in terms of additivity (which is well-defined only in the context of the analysis of variance). That is a technical issue which, though not lacking in interest, must wait for another occasion.

[7] For a fuller characterization of the notion of aggregativity, and the contrast with other more complex forms of systemic organization, see Richardson (1982).

[8] Again, Wimsatt speaks of *variance* in fitness that is *additive,* which I gloss as *variations* in fitness which are *aggregative.* I think Wimsatt's formulation is in error, but the differences will not matter for what we are doing here.

[9] I assume that a context-dependent effect at a given level could not reappear as an aggregative effect of context-independent variations at lower levels; in other words, I assume that a context-independent effect at one level will be context-independent at all higher levels.

[10] I confess, I find this somewhat less than obvious. The case after all is one which could as well be a single population where some individuals have the suppressor and some do not. In either this case, or the one they describe, there is no obvious error in thinking that we have selection against *XA* genotypes and none against *XS* genotypes (where the *XA* genotype is the genotype with the suppressor absent and *XS* is the genotype with the suppressor present). That is, there is no obvious flaw in embracing higher level units in such a case.

[11] Wimsatt is happy to concede that their differences are marginal at best.

[12] I do not deny that he engaged in speculations which could only be regarded as ventures into higher levels of selection (see, e.g., p. 87), but that does not deny the role of individualist thinking in his system.

[13] An extremely clear and useful introduction to the topic can be found in Roughgarden (1979), Part V. The discussion to follow relies heavily on Roughgarden, and especially on chapter 23.

[14] That is, we operate under the assumption of limited genetic diversity. Though the assumption is consonant with traditional population dynamics, and was once defended as an idealizing assumption on the ground that evolutionary response within populations is too slow to significantly affect ecological dynamics, it is generally regarded now as a false assumption.

[15] Specifically, within each subsystem, there is a tendency to maximize the equilibrium value for the predator population.

[16] High values for the influence of **A** on **a** or of **B** on **b** (which would correspond to high predation rates) will destabilize the subsystem, at least in cases of true predation. If they are sufficiently high, they may determine equilibrium values for extinction. Since this would correspond to a tendency to self-destruct, we will assume that the values here are relatively low. There is a moral to be had here for opponents of conservation.

[17] Interaction in all forms did not. It was common to press for a polygenic basis for traits, as can be seen in the controversies between Morgan and Castle.

[18] The dominance is actually partial. The hetozygous female has a diminished number of facets in comparison to the female homozygous for normal eyes, but the size of the overall surface is still considerably larger than that of the homozygous bar female. This effect was important in large part because it facilitated the use of this trait in genetic research.

[19] The change in name reflected his view that the double bar involved a duplication of a gene present in a single dose in bar individuals and lacking altogether in the individuals reverting to the 'wild type'.

[20] In Sturtevant's scheme, the homozygous bar had a single bar gene on each chromosome and the heterozygous double bar had two bar genes on one chromosome and none on the other. Hence, each had two bar genes, though unequally distributed.

[21] For an alternative approach to many of the same problems, which contends that considerations of the mode of organization alone cannot settle issues over the units of selection, see Richardson (1982).

[22] Though there are a number of very respectable theorists (both biologists and philosophers) who view kin selection as a form of group selection, the view seems to me mistaken. A full defense of the view that the unit of selection postulated in Hamilton's model is genic would be out of place here, but some indication of the sort of reasons which lead me to the view seem worth noting. First, there is no full calculation of individual fitness in the model. There is at most a schema for calculating the individual, and ultimately the extended, fitness contributions of a given trait. This in turn is taken to be determined by genic parameters, in particular the correlation determining the degree of genetic similarity between individuals. As far as the model goes, it is an open question what sort of function individual fitness is relative to its various components. It might be a simple aggregative function, or it might be complex enough that the individual must be treated as a unit of selection. The question is simply not raised. So, at least for the purposes of the model as developed, it is not the individual that is the relevant unit. Second, there is no *necessity* for any relevant group structure to facilitate kin selection. Selection will have a positive net effect if the benefits to kin outweigh the cost to the benefactor, even if this positive net effect is gained through a random distribution within a single panmictic community. There need not even be alternative groups, and hence need not be any group selection operating. Hence, the relevant unit cannot be the social group.

[23] Hamilton is quite clear that kin selection is a special case, and the crucial issue is only whether there is a genetic correlation between individuals. He explains "kinship should be considered just one way of getting positive regression of genotype in the recipient, and that it is this positive regression that is vitally necessary necessary for altruism. Thus the inclusive fitness concept is more general than 'kin selection'" (Hamilton, 1975, p. 201).

[24] Throughout, our interest has been in the level of adaptedness of organisms, or in their expected fitness, rather than their actual (or realized) fitness. Expected fitness is but one variable affecting actual reproductive success. For discussion, see Brandon (1978), Mills and Beatty (1979), and Burian (1983).

[25] For these issues, see John Beatty's groundbreaking (1980), and Maynard Smith (1978) as well as Richardson (forthcoming).

REFERENCES

Axelrod, R. and Hamilton, W. D.: 1981, 'The Evolution of Cooperation', *Science* **211**, 1390–1396.

Bateson, W.: 1913, *Problems of Genetics*, Yale, New Haven [reproduced, 1979].

Beatty, J.: 1980, 'Optimal-Design Models and the Strategy of Model Building in Evolutionary Biology', *Philosophy of Science* **47**, 532–561.

Brandon, R.: 1978, 'Adaptation and Evolutionary Theory', *Studies in History and Philosophy of Science* **9**, 181–206.

Burian, R.: 1983, 'Adaptation', in M. Grene (ed.), *Dimensions of Darwinism*, Cambridge University Press, Cambridge.

Bridges, C. B.: 1936, 'The Bar "Gene" a Duplication', *Science* **83**, 210–211.

Carlson, E. A.: 1966, *The Gene: A Critical History*, W. B. Saunders, Philadelphia and London.

Causey, R.: 1977, *Unity of Science*, D. Reidel, Dordrecht.

Darwin, C.: 1959, *On The Origin of Species.* A Facsimile of the First Edition with an Introduction by Ernst Mayr, Harvard University Press, Cambridge, 1964.

Dawkins, R.: 1976, *The Selfish Gene*, Oxford University Press, Oxford.

Dawkins, R.: 1982, *The Extended Phenotype: The Gene as the Unit of Selection*, W. H. Freeman, Oxford and San Francisco.

Dobzhansky, T.: 1932, 'The Baroid Mutation in Drosophila Melanogaster', *Genetics* **17**, 369–392.

Doolittle, W. F. and Sapienza, C.: 1980, 'Selfish Genes, The Phenotype Paradigm, and Genome Evolution', *Nature* **284**, 601–603.

Hamilton, W. D.: 1964, 'The Genetical Evolution of Social Behavior I and II', *Journal of Theoretical Biology* **7**, 1–16 and 17–52.

Hamilton, W. D.: 1975, 'Innate Social Aptitudes of Man: An Approach to Evolutionary Genetics', in R. Fox (ed.), *Biosocial Anthropology*. Malaby Press, London. Reprinted and abridged in R. Brandon and R. Burian (eds.), *Genes, Organisms, Populations: Controversies Over the Units of Selection*, MIT Press, Cambridge, 1984, pp. 193–202.

Hellman, G. and Thompson, F.: 1975, 'Physicalism, Ontology, Determinism, and Reduction', *The Journal of Philosophy* **72**, 551–564.

Levins, R.: 1974, 'The Qualitative Analysis of Partially Specified Systems', *Annals of the New York Academy of Science* **231**, 123–138.

Lewin, R.: 1982, 'Repeated DNA Still in Search of a Function', *Science* **217**, 612–623.

Lewontin, R. C.: 1970, 'The Units of Selection', *Annual Review of Ecology and Systematics* **1**, 1–18.

Lewontin, R. C.: 1974, *The Genetic Basis of Evolutionary Change*, Columbia University Press, New York.

Lewontin, R. C.: 1977, 'Fitness, Survival, and Optimality', in D. J. Horn, R. Mitchell, and G. R. Stairs (eds.), *Analysis of Ecological Systems*, State University Press, Columbus, OH, pp. 4–21.

Lewontin, R. C.: 1983, *Human Diversity*, W. H. Freeman, San Francisco.

Lewontin, R. C. and White, M. J.: 1960, 'Interaction Between Inversion Polymorphisms of Two Chromosome Pairs in the Grasshopper, *Moraba scurra*', *Evolution* **14**, 116–129.

Maynard Smith, J.: 1978, 'Optimisation Theory in Evolution', *Annual Review of Ecology and Systematics* **9**, 31–56.

Maynard Smith, J.: 1982, *Evolution and the Theory of Games*, Cambridge University Press, Cambridge.

Maynard Smith, J.: 1983, 'Game Theory and the Evolution of Cooperation', in D. S. Bendall (ed.), *Evolution from Molecules to Men*, Cambridge University Press, Cambridge, pp. 445–456.

Maynard Smith, J.: 1984, 'Game Theory and the Evolution of Behavior', *The Behavioral and Brain Sciences* 7, 95–125.

Mayr, E.: 1963, *Animal Species and Evolution*, University of Chicago Press, Chicago.

Mayr, E.: 1975, 'The Unity of the Genotype', *Biologisches Zentralblatt* **94**, pp. 377–388; reprinted with a recent preface in R. Brandon and R. Burian (eds.), *Genes, Organisms, and Populations*, MIT Press, Cambridge, 1984.

Mills, S. and Beatty, J.: 1979, 'The Propensity Interpretation of Fitness', *Philosophy of Science* **46**, 263–286.

Nagel, E.: 1961, *The Structure of Science*, Harcourt, Brace & World, New York.

Orgel, L. and Crick, F.: 1980, 'Selfish DNA: The Ultimate Parasite', *Nature* **284**, 604–607.

Richardson, R.: 1982, 'Grades of Organization and the Units of Selection Controversy', in P. D. Asquith and T. Nickles (eds.), *PSA 1982*, vol. 1, Philosophy of Science Association, pp. 324–340.

Richardson, R.: 1984, 'Biology and Ideology: The Interpenetration of Science and Politics', *Philosophy of Science* **50**: pp. 000–000.

Richardson, R.: (forthcoming), 'Models and Scientific Explanation', in a volume edited by G. Dorn and P. Weingartner.

Rosenberg, A.: 1978, 'The Supervenience of Biological Concepts', *Philosophy of Science* **46**, 263–288.

Rosenberg, A.: 1983, 'Fitness', *The Journal of Philosophy* **80**, 457–473.

Roughgarden, J.: 1979, *Theory of Population Genetics and Evolutionary Ecology: An Introduction*, MacMillan, London & New York.

Schaffner, K.: 1974, 'Reductionism in Biology: Prospects and Problems', in R. S. Cohen *et al.* (eds.), *PSA 1974*, D. Reidel, Dordrecht, 1976.

Sober, E.: 1980, 'Holism, Individualism, and the Units of Selection', in P. D. Asquith and R. N. Giere (eds.), *PSA 1980*, Philosophy of Science Association, 1981, pp. 93–121.

Sober, E. and Lewontin, R. C.: 1982, 'Artifact, Cause and Genic Selection', *Philosophy of Science* **49**, 157–180.

Sturtevant, A. H.: 1925, 'The Effects of Unequal Crossing Over at the Bar Locus in Drosophila', *Genetics* **10**, 117–147.

Sturtevant, A. H. and Morgan, T. H.: 1923, 'Reverse Mutation of the Bar Gene Correlated with Crossing Over', *Science* **57**, 746–747.

Sturtevant, A. H.: 1965, *A History of Genetics*, Harper & Row, New York.

Trivers, R.: 1971, 'The Evolution of Reciprocal Altruism', *The Quarterly Review of Biology* **46**, 35–57.

Williams, G. C.: 1966, *Adaptation and Natural Selection*, Princeton University Press, Princeton.

Wilson, E.: 1975, *Sociobiology: The New Synthesis*, Harvard University Press, Cambridge.

Wimsatt, W.: 1980a, 'Reductionist Research Strategies and Their Biases in the Units of Selection Controversy', in T. Nickles (ed.), *Scientific Discovery* vol. 2: *Case Studies*, D. Reidel, Dordrecht.
Wimsatt, W.: 1980b, 'Units of Selection and the Structure of the Multi-level Genome', in R. Giere and P. D. Aquith (eds.), *PSA 1980*, Philosophy of Science Association.
Wimsatt, W.: (in progress), 'VonBaer's Law of Development, Generative Entrenchment, and Scientific Change'.

ALEXANDER ROSENBERG

ADAPTATIONALIST IMPERATIVES AND PANGLOSSIAN PARADIGMS

ABSTRACT. This paper defends adaptationalist programs against criticisms offered in R. Lewontin and S. J. Gould, 'The Spandrels of San Marco and the Panglossian Paradigm.' It argues that defects Lewontin and Gould allege are essential features of a research program motivated by any theory which has the extremal character characteristic of the theory of natural selection. It suggests that the constraints on evolution to which Lewontin and Gould look as alteratives to adaptational mechanisms must themselves be adaptationally explained, unless these constraints are altogether non-biological. Gould's notion of 'exaptation' is shown to be consonant with and to provide the vehicle for the adaptational treatment of the developmental results of such constraints. To avoid "Panglossianism", adaptationalists need only be sophisticated in what they identify as phenotypes subject to selection.

Until the recent past, most of the history of biology was the pursuit of structures that would explain functions. Having pursued these structures to the natural stopping place of biochemistry and molecular biology, the discipline has neither gone further in its pursuit of structure to physical chemistry, and beyond it to physics, nor has it ceased at a dead end beyond which there are no further biological questions. Instead it has turned around, and now is actively engaged in the search for biological functions of chemically characterized structures.

This *volte face* is particularly apparent in the study of the genome. In the last decade molecular biologists have learned a great deal about the structure of the nucleic acid polymers that make up the chromosonal material. They have learned, in particular, that sequences of nucleic acids, the bases that compose DNA, adenine, thymine, guanine, and cytosine, recur in repeating series of several different kinds; that very similar sequences can have utterly different effects, that some have no apparent role in the genetic capacities of the DNA. Biochemists have learned that there are split genes, which contain "introns" – intervening sequences that do not code for any protein product, that are snipped out in the messenger RNA; that there are some so-called "moderately repeated" DNA sequences recurring hundreds of times on the same chromosome, and some "highly" repeated DNA-sequences instanced millions of times on the same chromosome; that some genes come in tandem clusters, and others are to be found on more than one chromosome;

James H. Fetzer (ed.), Sociobiology and Epistemology, 161–179.

that there are "psuedo-genes" with base sequences very close to those of functioning genes, but which do not have their structural or regulatory functions; then there are genes occurring only once on one chromosome, in each nucleus, that code for a protein-product in the way we expect genes to do. All this structural information cries out for functional explanation, and much of it for evolutionary, adaptationalist explanation. That is, while some features of the structure of the chromosome are what we might expect, given the role which the genome plays in the heredity and development of individual organisms, other features seem to do nothing for the organism that contains them. For these structures, a different sort of explanation is required, one which explains their presence and persistence, in spite of their neutrality or maladaptive contribution to the survival of the organism containing them. Typically, these explanations too will be functional, identifying features of structurally identified portions of the genome as contributing either to the evolutionary adaptation of the very portion of the DNA itself, or to the evolutionary adaptation of the population in which the organism interbreeds.

Adaptational explanation is as much in demand at the most fundamental level of biological theory as it is in ecology, or behavioral biology or sociobiology or physiology. This makes recent attacks on the propriety of such explanatory strategies particularly serious. For we cannot expect the need for such explanations to be superceded through the revolutionary developments of molecular biology. And when these attacks on adaptationalism are launched by the most methodologically concerned and philosophically well-informed biologists, the issue is of transcending seriousness.

I

Two such biologists, Richard I. Lewontin and Stephen Gould, have attacked general adaptationalist strategy that has animated the biological study of the genetic material, and in fact the whole evolutionary approach to explaining phenomena, and their origins. Their attack has been more influential than perhaps any other essay in the recent literature on the philosophy and methods of biology. They call the approach the "Panglossian" paradigm, recalling the optimistic doctor of Voltaire's *Candide*: Dr. Pangloss, who found a function subserved for every distinct effect of an event or structure we could identify. Just as Dr. Pangloss found that the nose has its shape to enable us to wear glasses, adaptationalists maintain, according to Lewontin and Gould, "the near omnipotence of natural selection in forging organic design and fashioning the best among possible worlds. This programme regards natural selection as so powerful and the constraints upon it so few that direct production of adaptation through its operation becomes the primary cause of nearly all organic form, function, and behavior. Contraints upon the pervasive power of natural selection are recognized of course. . . . But they are dismissed as unimportant, or else, and more frustratingly, simply acknowledged and then not taken to heart and invoked." (Lewontin and Gould, 1978, 584–585) The defect of adaptationalism, on Gould and Lewtonin's view, is that it makes false attributions of adaptive significance and therefore obscures the course of evolution, and that its ubiquity blinds biologists to alternative, better accounts of the origin and nature of biological phenomena. If they are correct, the adaptationalist strategy in molecular biology is the recapitulation of errors of method and theoretical presupposition common throughout biology.

The first step in the adaptationalist strategy to which Gould and Lewontin object is its division of organisms into traits, which are each to be separatly explained as the result of different evolutionary processes optimizing for different ends. Of course the identification of such traits, and their individuation, represent empirical hypotheses: guesses that what our nomenclautre distinguishes are in fact distinct and can respond to selection with some degree of independence. Often such guesses are wrong. Gould is well-known for having identified one such implicit hypothesis: the belief that the chin is a single trait, instead of the interactive result of two separate growth fields, the mandibular and the alveolar (Gould, 1977, p. 381).

The division of organisms into traits are conjectures about whether they are the natural kinds out of which organisms are constructed. Every biological study of the components of a system must begin with such fallible conjectures. No one can object to such "atomization," in principle, for without it the life-sciences could hardly proceed. But Lewontin and Gould seem to suggest that adaptationalists always or for the most part get it wrong, that their segmentation of organisms into parts each with their own distinct evolutionary path, is wrong. Often these erroneous divisions are revealed as such in the failures of attempts to give them evolutionary explanations. Lewontin and Gould seem to recognize this:

> After the failure of part-by-part optimization, interaction is acknowledged via the dictim that an organism cannot optimize each part without imposing expenses on the others. The notion of 'trade-off' is introduced, and organisms are interpreted as best compromises among competing demands. Thus, interaction among parts is retained completely with the adaptationalist programme. Any suboptimality of a part is explained as its contribution to the best possible design for the whole.

This is why a "story" can always be told that explains any possible increase or decrease of a hereditarily linked sub-species by appeal to selective forces *known or unknown*. In other words, the pervasive character of extremal theories is but the other side of the coin from their insulation against falsification.

Now, all theories are strictly unfalsifiable, simply because testing them involves the employment of auxiliary hypotheses. But extremal theories are not only insulated against strict falsification, they are also insulated against the sort of falsification that usually leads to modification of theories instead of auxiliary hypotheses. In the case of a nonextremal theory, falsification may lead either to revision of the description of test conditions, or to revision of the theory by the addition, for example, of new antecedent conditions to its generalizations, or new qualifications to its *ceteris paribus* clauses. But this is not possible in the case of extremal theories. The assumptions of theories like Newton's, or Darwin's, do not embody even implicit *ceteris paribus* clauses. The latter does not, for example, assume that selection maximizes fitness *ceteris paribus*. With these theories the choice is always between rejecting the auxiliary hypotheses, the description of test conditions, or rejecting the theory altogether. For the only change that can be made to the theory is to deny that its subjects invariably maximize or minimize its chosen variable. This, of course, explains why high level extremal theories like Newtonian mechanics are left untouched by apparent counter-instances and are superseded, not by qualified versions, in which antecedent conditions are added, but by utterly new extremal theories, in which the values of very different variables are maximized, or minimized, or by new theories of a non-extremal type.

Extremal theories have become an important methodological strategy because of the success of the earliest of them, Newtonian mechanics, but also because their insulation to falsification has enabled them to function at the core of research programs, turning what otherwise might be anomalies and counter instances into puzzles, opportunities for extending the domain and deepening the precision of the extremal theory. But a necessary condition for the success of extremal theories and their associated research programs is that there be at least some guide to or agreement on independent specifications, characterizations, or ways of identifying the determinants of their subject's extremal behavior. These guides are themselves theoretical claims. But it is essential that they are distinct from the theory's claims about its subject's maximizing or minimizing behavior. An example from mechanics will make this point clear. Treating the Newtonian force law, $F = ma$, as a synthetic proposition governing the behavior of bodies, requires that an independent way of specifying mass, *m*, be proved. Hooke's spring law provides such a specification. This independent specification not only preserves the contingent character of the force law, but enables us to employ it to explain the operation of springs, projectiles, fluids, and much else that mechanics account for. It is a well-known criticism of the extremal claims of the theory of evolution that they are empty tautologies because the theory provides no independent specification for the concept of fitness, the variable whose rate of change is maximized within evolving populations. Since higher levels of fitness are cited by the theory to explaon greater reproduction rates among populations, these same rates of reproduction cannot provide the independent specification of different degrees of fitness. Accordingly, unless such an independent specification is available, the theory is guilty as charged of being empirically empty. Much of the literature in the philosophy of biology has been devoted to discussing the difficulty and the possibility of providing such an indepndent specification. (See for instance Williams, 1973; Mills and Beatty, 1979; and Rosenberg, 1983).

Thus, Lewontin and Gould's criticism of the adaptationalist program reflect real and unavoidable features of the theory which animates it: the refutation of one adaptionalist "story" does result in the provision of another; the failure to find any is considered a scientist's failure, not a theoretical falsification; and the criteria for acceptance of adaptational "stories" are very loose, a result of the very nature of the theory they reflect. But the theory they reflect is the theory of natural selection, and neither Lewontin nor Gould want to surrender this biological achievement itself (though Gould at least has rejected a view he calls "Darwinian evolution," the claim that evolution proceeds by continual selection over small variations at a constant rate). Accordingly, their methodological criticisms reflect a misunderstanding of the theoretical exigencies of an extremal theory, in particular, and of the fact that no theory can be strictly falsified, in general. There is nothing methodologically wrong with constructing a new adaptationalist explanation of a given "form, function of behavior" when an older one has been rejected; indeed, the theory of natural selection requires it in virtue of its extremal character.

There is, however, an ancillary and dispensable hypothesis, often uncritically employed in the application of the adaptationalist program, that is open to criticism. And the criticism may encourage the "developmental" alternative Lewontin and Gould favor. This doubtful assumption is the hypothesis that the items whose persistence is to be explained as adaptations are those individuated and identified in ordinary language or in one or another biological theory of relatively restricted generality. Lewontin and Gould cannot object to the division of organisms into independent traits, each of which can separately respond to selective forces. Their objections must be to the particular divisions adopted by evolutionary biologists. To see this we need note only that Lewontin and Gould advocate or at least permit adaptive theorizing about "integrated packages" within which components constrain the development and presumably the evolution of one another. As they note, "differentiation of organ systems, and their integration into a functioning body is . . . a delicate process. . . . development occurs in integrated packages, and cannot be pulled apart piece by piece in evolution." Thus within a package, any one component evolves under the constraint of other components (and of its own developmental precursors) with which it must remain not merely compatible, but functionally cooperative. Similarly, whole packages are equally constrained by the others together with which they compose the whole organism. Accordingly we cannot assign adaptations and explanations based upon them to components and to packages which do not reflect the constraints under which they operate and evolve. These constraints may be so great that they overwhelm even moderately strong selective forces to which individual components and packages would respond were they independent. As a result there might be much less evolutionary adaptation of individual components, packages and organisms even during long periods of considerable environmental and genetic variation. From this Lewontin and Gould infer that "the adaptational programme cannot explain the alterations of developmental programmes underlying nearly all changes in *Bauplane*." But this will only be true of adaptationalism if it is not allowed to apply its explanatory scheme to the developmental packages and is not allowed to include among the selective forces shaping any one component the constraints of the rest of the "package" in which it participates.

Lewontin's and Gould conclusion only follows if adaptationalists are committed to the particular taxonomy of selectively independent traits which they have inherited or uncritically chosen or which their opponents have foisted upon them. Such uncritical adaptationalists differ from Gould and Lewontin not on a point of biological methodology, still less on a matter of general theory, but on a factual question of how organisms are to be carved up into parts which face selective forces as units. Freed from a superficial typology of traits, which divides organisms up into as many parts as we have names for, the adaptationalist, i.e., the evolutionist, can avail himself of each of the five explanatory approaches Lewontin and Gould sketched as potential alternatives of his "program". Only if Lewontin and Gould hold the radically holist view that organisms cannot be broken into components at all, are "not decomposable into independent and separately optimized parts", can we find anything in their attack on Panglossianism with which the adaptionalist cannot, nay, should not reconcile himself. But though they seem to endorse such a radical holism for these purposes, in their actual biological and particularly genetical and paleontological practice, both repudiate any such commitment.

But it may be replied that these concessions to Lewontin and Gould are nothing less than the surrender of adaptationalism, for they seem to require surrender of the thesis that "selection can break any correlation and optimize parts separately." In fact they do not; these admissions require greater sophistication about what a "part" is, and what distinguishes components of a single unit from correlated but independent units. To see this consider how we identify units that respond to selective forces and how we distinguish correlations from components by an adaptationalist approach. If the adaptationalist explanation for what is individuated as an organ or a part or a bit of behavior fails, this is one sign that we have not individuated the components of the organism right. Explanatory failure suggests that what we have focused on is perhaps not a "developmental package" but an artificial jerrymandering of the organism. Or again, suppose we hypothesize that two features are only correlated for nonevolutionary causes, but we can find no reasonable independent adaptationalist explanations for each of them. One alternative is to deny that they are merely correlated, and treat them as components of a "package" that evolves as a unit. In fact, the only way to reach Lewontin and Gould's aim, "a pluralistic view [that] could put organisms with all their recalcitrant, yet intelligible, complexity back into evolutionary theory" [p. 597], is by pursuing an adaptationalist program.

The adaptationalist program is just the one that must be endorsed, if the only general forces that have operated to arrange each of the levels of biological organization are those of natural selection operating on successively greater aggregations of macromolecules. Such operations would generate just the "developmental packages" which Lewontin and Gould urge us to make the objects of biological explanation. If these are the only general forces which have resulted in this hierarchy, then the only general theories to be uncovered in biology are the theories of natural selection and of molecular biology (to the extent it may be distinguished from organic chemistry). If these theories exhaust the theoretical possibilities and if research programs must reflect underwriting theories, then the only viable research program in biology is the adaptationalist one. The only alternative is to hold that there are distinct regularities and theories discoverable in biology intermediate between chemistry and selection. It is of course possible that Lewontin and Gould believe there to be such autonomous generalizations, say, in developmental morphology or phylogeny, generalizations with enough content, precision and confirmation to be deemed laws, with enough coherence to constitute a theory, and to animate a research program which may really compete with adaptationalism. The only way to undermine adaptationalist claims on biology would be to provide such a theory.

In a particularly important paper, Stuart Kauffman (1983) reports studies which may eventuate in such theories, expressing laws, or as he expresses it "biological universals" intermediate between those of biochemistry and the principles of natural selection. As Kaufmann notes, the ultimate source of order in the living world must derive from basic properties of structural and dynamic self-organizing properties of complex molecular systems, which provide and which limit the possibilities explored by variation, and selection. There is nothing particularly mysterious about the "self-organization" Kaufmann appeals to. Its simplest manifestation is the aggregation and layering of lipids we can detect when oil is mixed with water and allowed to stand. This "self-organization" is a consequence of molecular interaction, but will obtain among lipids no matter what their particular molecular size or composition. Similarly, other properties of macromolecular assemblies may appear just through stochastic forces or thermodynamic process, or as expressions of topological features of surfaces, no matter what their composition. Instances of such "universals" may recur at independent nodes in evolutionary phylogeny, not as a result of convergent adaptational evolution (though they may lead to it), but because at these nodes the purely physical initial conditions of the appearance of these biological universals obtain. These properties will serve as the building blocks on which selection operates, and their occurrence also constitute constraints that selection cannot circumvent, any more than it can circumvent gravity. To the extent that such self-organizing properties can be systematized and predicted from purely physical and chemical considerations, they do not constitute a source of the sort of autonomous generalizations about development that Lewontin and Gould look to for an alternative to the adaptationalist approach in the explanation of biological phenomena. But to the extent they are significant constraints and or a substantial part of the material evolution must operate on, they will provide much of the explanatory resources in biology. If what Kauffmann calls biological universals cannot be explained by physical principles or by evolutionary ones, then a theory that treats of them may provide the kind of alternative to adaptationism which Gould and Lewontin look to.

III

Gould has himself offered a terminological suggestion which may help clarify the adaptationalist offered a terminological suggestion which may help clarify the adaptationalist controversy. Adaptationalism must be compatible with the possibility, indeed the actuality, that many evolved structures have a present function, but were "not built by natural selection for their current roles" (Gould and Vrba, 1982, p. 6). In such cases the identification of current role is no part of the evolutionary explanation of the origin and persistence of the structure. What is more challenging to the adaptationalist, such structures may not have arisen through any adaptive mechanism.

Gould and Vrba illustrate this problem by appeal to several arresting examples, among them the problem of assigning a function and constructing an adaptationalist explanation for highly repetitive DNA. Their first example is the evolution of birds and the appearance of feathers. The earliest known birds were fully feathered, and yet capable of only very limited flight. Accordingly, the notion that feathers have their origin in the adaptation they provide for flight seems implausable. Alternatively, it has been suggested that feathers first appeared as an adaptation for thermo-regulation and later found a new role, unconnected with thermo-regulation in the trapping of insects, by increasing hand-surface area. Only later, together with other evolutionary changes, did they find a role in flight. By this point further changes in the nature of bird-feathers may have come directly under the influence of selection for flight, but, on this theory, at any rate, its origins have nothing to do with adaptation for its present role.

After sketching similar accounts for the origins of bone and of an essential mammalian enzyme, Gould and Vrba turn to the puzzle of identifying adaptive functions and tracing evolutionary histories of portions of the repetitive genetic material described above:

> We are now, not even twenty years [after the decoding of DNA] faced with genes in pieces, complex hierarchies of regulation, and above all, vast amounts of repetitive DNA. Highly repetitive of satellite, DNA can exist in millions of copies; middle repetitive DNA, with its tens to hundreds of copies, forms about one quarter of the genome in both *Drosophila* and *Homo*. What is all this repetitive DNA for (if anything)? How did it get there? (p. 10).

Two explanations that Gould and Vrba treat as inadequate are, first, that repetitive DNA plays an immediate regulatory role in cell-division, a role for which it was selected, and, second, that such DNA has had a role in freeing genes to vary and to reshuffle in ways not bound by narrow selective forces. They suggests a third alternative:

> [P]erhaps repeated copies can originate for no adaptive reason that concerns the traditional Darwinian level of phenotypic advantage. ... Some DNA elements are transposable; if they can duplicate and move, what is to stop their accumulation as long as they remain invisible to the phenotype (if they become so numerous that they begin to exert an energetic constraint then natural selection will eliminate them)? Such "selfish DNA" may be playing its own evolutionary game at a genic level, but it represents a true nonadaptation at the level of the phenotype. Thus, repeated DNA may often arise as a nonadaptation. Such a statement in no way argues against its vital importance for evolutionary futures [i.e., for subsequent phenotypical adaptation of repeated DNA]. When used to great advantage in that future, these repeated copies are *exaptations* (p. 10, emphasis added).

Exaptation is Gould's and Vrba's neologism, one introduced to enable us to distinguish structures whose current functional roles are no part of the explanation of their origin from those whose roles are. Their terminological recommendation is as follows. "Adaptation" should be reserved for a character with a current use, role or function in meeting an organism's needs, if the character's origin hinges on selection for that role or function. An "exaptation" is a character with a current role, or function, but whose origin is either the result of selection for a different role or function, or else the result of some non-adaptational process altogether. Gould suggests that both adaptations and exaptations come under the heading of "Aptation", where this term describes biological items that are normally described as having functions or, at least, as having biologically significant effects. Thus, feathers are exaptations because they have been co-opted for flight, though their adaptive origin is in thermo-regulation. 'Selfish DNA' may now be an exaptation for increasing the rate of evolutionary variation, or for facilitating genetic recombination, though it arose through no adaptive process whatever. An exaptation is, according to Gould and Vrba, what has hitherto been called a pre-adaptation. The notion of pre-adaptation has been employed among biologists with some discomfort. For the term suggests an immanent directedness of evolution towards some as yet unrealized adaptation. The label pre-adaptation is given to a trait retrospectively, after an 'aptation' has become an adaptation. Such an aptation is, on these definitions, an exaptation.

This terminological suggestion of Gould's seems to afford scope in evolutionary studies for the anti-adaptationalist alternatives that he and Lewontin have advanced and for Gould's well-known hypothesis of punctuated equilibrium as the "mode and tempo" of evolution (see Gould in Milkman (1982) for an introduction to this hypothesis). For he notes that

> Features co-opted as exaptations have two possible previous statuses. They may have been adaptations for another function, or they may have been nonaptive structures. The first has long been recognized as important, the second underplayed. Yet the enormous pool of nonaptations must be the wellspring and reservoir of most evolutionary flexibility (12).

The second alternative is that features or structures have arisen and persisted for nonadaptional causes, that their character is a function of the constrains provided by the organization in which they are found. These constraints fix traits and characters over geologically long periods and are only broken in a way that permits adaptive diversity on limited occasions, occasions which reflect bursts of evolution that punctuate the otherwise changeless sequence of terrestrial history. It is only on these occasions that adaptation may come into play, revealing exaptations otherwise hidden from selective forces.

As we have seen above, this conclusion can itself result from adaptationalist research, once the right units that face selection are uncovered. This much is evident in Gould's and Vrba's own description of the potential exaptational character of 'selfish DNA.' They note that "repeated DNA can originate for no adaptive reason that concerns the traditional Darwinian level of phenotypic advantage." But, of course, adaptationalist hypotheses are not limited only to the explanation of what can be described in the terminology of "traditional Darwinian phenotypes." Freed from this unreasonable restriction, the adaptationalist can help himself to Gould's own claim that "selfish DNA" may be playing its own Darwinian game at the genic level. Gould's influence that "it represents a true nonaptation at the level of the phenotype" does not follow if we are permitted to adopt a relatively sophisticated notion of the phenotype and of the items whose evolution is governed by the theory of natural selection (R. Dawkins develops just this sort of a notion of phenotype in *The Extended Phenotype* (1981)). Since one of the interpretations of that theory makes it a theory about the evolution of segments of the genetic material, it empowers the biologists to pursue an adaptationalist approach to molecular evolution. The identification of features as exaptations is no reason to surrender this program. Indeed, it enables us to apply the theory even where, as in this case, adaptationalist hypotheses at higher levels of organization must be viewed with suspicion.

Furthermore, it becomes apparent that adaptationalism, and its motivating theory, is perfectly compatible with – indeed might lead us to expect – punctuated equilibrium in evolution if, as Gould and Lewontin hold, the components of organisms and their parts are the chief evolutionary constraints on one another's responses to selection. For, if this is the case, the fitness of any one part and any one "package" supervenes heavily on other parts and "packages" with which it is grouped. By channelling other selective forces, those of the environment, into relatively narrow streams, these constraints will result in long periods of apparent stasis, periods during which almost any variation will prove inviable because it breaks up the developmental programs required for life. Only rarely will massive environmental changes or fortuitous variations result in new forms that are viable. When it does, these new forms will have to differ very markedly from their predecessors, just because they break up significant constraints successfully. This is just what punctuated equilibrium demands. But it appears to be entirely the product of a multi-leveled adaptationalist mechanism which does nothing more than justice to all the selective forces governing evolution. (Of course, the consistency of a punctuated equilibrium approach with an adaptational one must not be misunderstood. It does not constitute an endorsement of the former view, only another argument to show that the dispute between gradualism and punctuated equilibrium is one within the ambit of the theory of natural selection, despite the lable of "Darwinian Evolution" which one party to the dispute employs to describe the thesis of its opponents.)

With this reconciliation in mind, we may usefully employ Gould's distinction between adaptations and exaptation to make clear the sorts of functional questions molecular biologists want to ask about the variated structure of the eucaryotic genome. Some of their questions are about immediate function: for example, what physiological needs of their hosts and of themselves are served by quantities of highly repeated DNA? The answer to this question will probably be different from the answer to the question of what selective force resulted in their origin and persistence. We can mark this difference by calling them exaptations, thus distinguishing between the questions that biologists raise about them.

IV

But we cannot do without the questions that the agenda of adaptationalism sets. At least we cannot so long as biology continues to pursue distinctively biological questions, questions about function which physical science cannot answer and whose provenance it cannot deny. For biological systems are composed of subsystems that have functions. Not all of these functions have direct evolutionary explanations, but many of them do, and all must eventually reflect the operation of natural selection on variation somewhere in their causal ancestry. To this extent adaptationalism will always be with us. Indeed, it is hard to see what biology would be like without it. For there is really no alternative to adaptationalism – to the Panglossian paradigm – short of surrendering the theory of evolution itself. We might, of course, surrender what Gould calls Darwinian evolution, the thesis that natural selection proceeds at a uniform rate through small variations, but this is a thesis about the values of evolutionary parameters and the nature of the underlying mechanism. It is not the heart of evolutionary theory; surrendering it does not touch the notions of variation and selection as the mechanism of evolution when it occurs.

In its full dress, Gould and Lewontin's attack on adaptationalism can only be taken seriously as an attack on natural selection itself. Some of Lewontin's discussions of this theory do in fact encourage such an interpretation (cf. for instance Lewontin, 1978). A more limited target, however, suggests itself for their attack: The unreflective optimistic predelictions of evolutionary biologists and the hidden ideological agendas of sociobiologists (cf. Lewontin, 1980). But if the real aim of their attack is to undercut the expectations of optimists and the credibility of ideologues, then Lewontin and Gould have been guilty of over-kill, of throwing out the baby with the bath-water, of slaughtering the goose that lays the golden eggs. For they have savaged evolutionary theory itself and not the disreputable abuse of it they may have hoped to further forestall.

There remains one important lesson to draw from their paper. Evolutionary biology should certainly allow for the developmental constraints they declare to be so pervasive in the evolutionary process. Indeed, it should do more than merely allow for them; it should help explain them. It can only do this if the theory and its associated method are freed from any particular parochial or restricted way of describing the units which face selective forces. Just because biology or ordinary language has a name for some part of a biological system is no reason to think the item named is an independent, distinct unit, whose function or persistence needs to be or can be explained. How we divide up biological phenomena into their units is not a pretheoretical matter, to be decided before we turn to their evolutionary explanation. Deciding on what are the facts to be explained goes hand in hand with formulating the best explanations of the facts. This is a methodological maxim that biologists must keep firmly in mind. And no one should keep it more firmly before them than molecular biologists and sociobiologists. The former need to continually reflect that the structural units which their chemical methods reveal to them in the study of the genome may not be the units which faced selection, that combinations of nucleic and sequences whose evolutionary unity is not revealed by any combination of procedures from organic or physical chemistry, may nevertheless have an adaptive explanation together, though not separately. Sociobiologists need to keep this possibility even more firmly in mind. For the phenomena they deal with are already divided up by ordinary language of ethological observation and, in the human case, by the typology of "folk psychology" (in Churchland's sense; cf. Churchland, 1981). It is tempting to find adaptationalist explanations for long-lived institutions and persistent behaviors described in ordinary language. Indeed, much of the outrage to which sociobiologists have been exposed is predicated on the supposition that it is social arrangements as we ordinarily describe them that are to be identified as selectively fixed, "biologically determined." It is a mistake made equally by sociobiologists and by their detractors to suppose that their theory hinges on finding adaptationalist explanations for human behavior classified in the type-vocabulary of ordinary behavior. It will be ironical if reflection on Lewontin and Gould's attack on adaptationalism frees the sociobiologist from this mistaken supposition. For, in doing so, they will have provided sociobiology with its best chance of success.

Syracuse University

REFERENCES

Churchland, P. 1981, 'Eliminative Materialism and the Propositional Attitudes', *Journal of Philosophy* **78**, 67–90.

Dawkins, R.: 1981, *The Extended Phenotype*, Oxford University Press, Oxford.

Gould, S. J.: 1977, *Ontogeny and Phylogeny*, Harvard University Press, Cambridge.

Gould, S. J. and Vrba E.: 1982, 'Exaption – A Missing Term in the Science of Form', *Paleobiology* **8**, 4–15.

Kauffman, Stuart: 1983, 'Filming Some Epistemological Gaps', *PSA 1982*, pp. 292–313.

Lewontin, R. and Gould, S. J.: 1978, 'The Spandrels San Marco and the Pangolossian Paradigm: A Critique of the Adaptationalist Programme', *Proceedings of Royal Society of London* **205** (1978), 581–598.

Lewontin, R.: 1978, 'Adaptation', *Scientific American*, September.
Lewontin, R.: 1980, 'Adaptation', *Encyclopedia Einaudi*, Milan.
Milkman, R.: 1982, *Experimental Population Genetics*, Van Nostrand Reinhold, New York.
Mills, S. and Beatty, J.: 1979, 'The Propensity Definition of Fitness', *Philosophy of Science* **46**, 263–286.
Rosenberg, A.: 1983, 'Fitness', *The Journal of Philosophy* **80**, 457–473.
Williams, M.: 1973, 'The Logical Status of Natural Selection and Other Evolutionary Controversies', In M. Bunge (ed.), *Methodological Unity and Science*, D. Reidel, Dordrecht, pp. 84–102.

ELLIOTT SOBER

METHODOLOGICAL BEHAVIORISM, EVOLUTION, AND GAME THEORY

ABSTRACT. A review of some standard philosophical criticisms of methodological behaviorism shows that the only substantive challenge to that position allows that there may be research programs in which behaviorism is an appropriate methodological stance. Evolutionary biology provides some cases in point. Behaviorist assumptions are at work in sociobiological arguments, in some models of cultural evolution, and in game theory. The defects that there may be in these approaches do not include the behaviorist assumption. The perspective developed throws light on the charge that sociobiology uses anthropomorphic descriptors of the traits it investigates.

1. ROCK, PAPER, SCISSORS [1]

It is perhaps not an entirely idiosyncratic picture of the evolution of views on the mind-body problem during the last thirty years to see in it a progression from behaviorism to the mind-brain identity theory to functionalism. Each of these positions pinpointed defects in the one it supplanted. The identity theory (Smart, 1959; Place, 1962) saw that there can be no *a priori* objection to postulating internal mental states; functionalism (Putnam, 1967; Fodor, 1968, 1975) saw that the mental properties discussed in psychological theory need not be type-identical with physical properties of the central nervous system.

This picture of progress has its merits. But its significance has, at least in one respect, been misconstrued. If there is one view that philosophers now generally believe to have been rightly deposited on the rubbish heap of history, it is methodological behaviorism. I want to argue that there are influential areas of investigation in evolutionary theory that presuppose the truth of methodological behaviorism. One involves the kinds of hypotheses about genetic evolution put forward by sociobiologists. Another involves the development of models that attempt to describe commonalities between genetic and cultural evolution. Whether either of these research programs will be fruitful is currently a matter of controversy. My point here will be that their adherence to methodological behaviorism is not in itself a defect, but merely follows the lead of a longstanding pattern of explanation deployed in evolutionary theory.

My defense of methodological behaviorism in these contexts mimics some

James H. Fetzer (ed.), Sociobiology and Epistemology, 181–200.

of the reasons that functionalism produced in its criticism of the identity theory. The progression from behaviorism to identity theory to functionalism allows us to turn full circle and see what behaviorism has to offer after all.

2. WHAT IS WRONG WITH METHODOLOGICAL BEHAVIORISM?

Philosophers have had a field day with methodological behaviorism largely because its defenders produced manifestly defective arguments on its behalf. But it is one thing to refute a thesis, quite another to show that some argument for it is flawed. Both undertakings have their point, of course, but we now, in retrospect, need to get clear on where things stand and why.

I take methodological behaviorism to be the thesis that inner states need not be postulated to explain behavior. Taken strictly, it is a methodological thesis, not an ontological one. The claim is not that there are no inner goings on, but just that they need not be invoked. Of course, when linked to a principle of parsimony, the methodological idea may lead to an ontological conclusion. If mental states need not be postulated to explain *anything*, one might take this to be a reason to think that there are no such things. No one ever thought to pursue this line of reasoning with respect to inner *physiological* states. These we have excellent reasons to believe in, whether or not they matter to psychology. But methodological behaviorists often seem to have taken the further, ontological, step with respect to the mental. An implicit premise was that the explanation of behavior is the only possible justification there can be for claiming that mental states exist. The testimony of introspection was ruled out of court; the idea that we can have reason to think that something exists without its explaining anything was never really considered.

Notice that methodological behaviorism and so-called logical behaviorism are almost opposite in their implications. I am interested only in the former. Methodological behaviorists assumed that beliefs and desires are inner states and decided to ignore them. Parsimony then led naturally to the conclusion that such items do not exist. Logical behaviorists, on the other hand, denied that beliefs and desires are inner states and identified them with items of behavior. By no stretch of the imagination can a logical behaviorist consistently deny that people have beliefs and desires, if the having of beliefs and desires simply consists in exhibiting various behavioral regularities. Logical behaviorists, after all, do not doubt that behavioral regularities *exist*.

Anyhow, getting back to the proferred refutations of methodological behaviorism, what are we to find? I take as my text here the fine article by

Dennett (1978) called 'Skinner Skinned.' In that essay, Dennett does us the favor of assembling the various remarks that Skinner has made in defense of his position. Skinner objects to mental states on the grounds that they are unobservable and inner (although he occasionally denies that this is the real basis for his animus against the mental). Dennett quite rightly points out that such objections go contrary to familiar principles of scientific method. My point here is that this criticism of methodological behaviorism succeeds only in undermining one argument on its behalf; it does not show that the view is false.

Similar remarks apply to Dennett's assessment of Skinner's charge that mentalism makes explanation "too easy." According to Skinner, the trouble with invoking beliefs and desires to explain behavior is that a mentalistic story of this sort can always be concocted, no matter what behavior one wishes to explain. Denntt's reply is that the same can be said of the behaviorist's preferred mode of explanation in terms of stimulus-response.

Skinner and Dennett are right about one thing, but wrong about another. The bald claim that I'm writing this page because of my beliefs and desires, or because of my stimulus-response history, is true but rather uninformative. Indeed, if I were not writing this page, but were playing baseball instead, *that* fact would be explained by my beliefs and desires and by my stimulus-response history. The belief-desire strategy and the stimulus-response strategy are *generally* applicable, according to their practitioners; regardless of which of a wide range of behaviors a subject produces, the requisite form of explanation is in principle available. However, the same flexibility cannot be said to attach to *particular* explanations that follow either strategy. The particular constellation of beliefs and desires one might invoke to explain why I am now writing this page would *not* explain my playing baseball; similarly for the stimulus-response history that might be invoked for my writing. *General strategies* of explanation work for a wide range of alternative behaviors; *particular explanations* do not have this sort of flexibility.[2]

Once again notice where Dennett's objection leads. Methodological behaviorism, says Dennett, has failed to identify a genuine defect in mentalism. This at most shows that the behaviorist *argument* fails; it does not refute the behaviorist *thesis*. As before, Dennett's argument succeeds in removing an impediment to mentalism – there is no in principle objection to postulating mental states. But possibility is not necessity. We have yet to see why, contrary to behaviorism, mental states are *needed* to explain.

Dennett deploys a third line of reasoning, one that does a better job of refuting behaviorism. Behaviorism cannot explain so-called "creative" behaviors – ones that are unprecedented, relative to the subject's stimulus-response history. Dennett discusses Skinner's example of a person's handing over his money to a robber, when the robber says "your money or your life." Even someone who has never been robbed knows what to do; this prudent behavior, says Dennett, is beyond behaviorism's ken.

Dennett does not fill in the details very much as to how this argument is to work. The behaviorist will insist, presumably, that the subject's stimulus-response history must have conditioned him to hand over his wallet in that circumstance. Dennett sees this as an empty claim – it has the defect of being "too easy." The behaviorist will view his program's inability to fill in the details as a temporary impediment, one that future research and theorizing will remove. Dennett does not take the further step of showing why this reply is mistaken – why behaviorism's present inability to explain so-called "creative" behavor is an in principle limitation of that approach.

There is a fairly definitive criticism of methodological behavorism that can be teased out here, one that is fairly familiar. My interest in laying it out now is to make explicit the *kind* of criticism it is. The pattern of argument is something we will have occasion to reflect on later.

Everyone – behaviorist and nonbehaviorist alike – ought to agree that when I hand over my wallet, there is *something* about the stimulus situation that causes me to do this. By the same token, everyone ought to agree that there is something about my internal state at the time that makes me "receptive" to the robber's causal influence.[3] Rocks, after all, don't hand over wallets when threatened. So there are two questions that clearly are on the explanatory agenda.

The substantive commitment of behaviorism consists in how that view proposes to answer the first question. What is it about the stimulus situation that causes the behavioral response? It will not do, of course, for behaviorists to use mentalistic terminology in answering. They cannot explain my behavior by saying that the stimulus situation is such that I believe I'll be shot if I don't hand over my cash. A properly behaviorist answer must characterize the stimulus situation in nonmentalistic terminology.[4]

For some behaviors – including some that we might regard as "uncreative" – there is a physicalistic description of the stimulus that does the trick. When a male stickleback attacks a conspecific that comes within his territory, what is it about the stimulus situation that makes him do so? The answer, it appears, is the intruder's *red belly* (Tinbergen, 1951). The question for behaviorism is whether all behavior can be treated in this way.

Actually we must refine the challenge to behaviorism a bit more. The behaviorist is obliged to provide a physicalistic description of the present stimulus situation, a physicalistic description of the previous ones in which the "conditioning" took place, *and* a relationship of *physical* similarity between the past history and the present event.

This conjunctive requirement is essential if we are to capture the real thrust of the behaviorist idea. I don't think one can deny that there are various *physical* properties of the present stimulus situation that cause me to hand over my wallet. Nor do I think one can deny that the same is true for each of my past behaviors. In each instance, there are physical properties of the stimulus situation that are causally efficacious. Although there are views of causality that deny this,[5] I take this concession simply to reflect a kind of minimal and plausible physicalism.[6]

The problem for behaviorism comes when the view is asked to say how previous stimulus situations – the ones that allegedly shaped my current disposition to respond – are related to the present one. Both past and present have their physical descriptions. The question is what these descriptions have in common. Even if I have been threatened before and was "conditioned" from these experiences to respond by making concessions, this does not vindicate the behaviorist approach. What must be shown is that the physical circumstances of my past conditioning were the same – *in some interesting physical sense of "the same"* – as the present case. Only then will we have assimilated the human behavior to the stickleback case.

Another way to bring out this idea is to ask the behaviorist to give an explanation – not of a single token event – but of what a set of token events have in common. Think of the various physical techniques that a robber might use to cause me to hand over my wallet. These will be physically quite heterogeneous. Each will be such that there is a physical description of the stimulus situation that explains why I hand over my wallet. However, I doubt that there is a physical description of what those situations have in common that does the same.

It is at this point that mentalism leaps into the breech. A mentalistic explanation may postulate an internal representational state that allows us to say that in all these stimulus situations there is some uniform fact that does the causing. Although the stimulus situations differ physically, in each case they are related in the same way to the subject's mental state. They each produce, let us say, a fear of danger and the belief that handing over one's wallet is the way to avoid it. Of course, the hard part of mentalism is describing how environmental stimuli can produce such internal states. However, my point here is not to highlight pitfalls but to point to a pay-off.

By bringing the internal state of the subject into play, mentalism delivers a kind of general perspective that behaviorism cannot provide (Sober, 1983).

The reader familiar with the literature on functionalism will have anticipated how functionalism's critique of the mind/brain identity theory parallels mentalism's critique of methodological behaviorism. Again, the crucial concept is *generality*. Neither identity theorists nor their functionalist critics should deny the fact that a given token behavior has a physicalistic explanation in terms of states of the subject's central nervous system.[7] It should not be controversial that some story about the brain explains why I hand over my wallet in the robbery. This and other psychologically similar behaviors are each explicable in terms of the subject's brain state. However, it is when we try to unify these various examples within a common framework – for example, by saying what perceiving the presence of danger amounts to – that the functionalist sees the identity theory as falling into difficulty. For functionalism maintains that psychological states may have many different physical realizations. The particular brain configuration that allows me to see that I am in danger is only one of many devices that could do the trick. According to functionalism, it is the explanation of similarities among token events, and not explaining token events one by one, that reveals the weakness of the identity theory.

3. WHAT IS RIGHT ABOUT METHODOLOGICAL BEHAVIORISM?

Mentalism faults behaviorism for missing significant generalizations. Within the purview of mentalism, functionalism faults the identity theory for the same thing. It now is time to turn full circle and circumscribe the kinds of generalization that behaviorism can hope to find but which mentalism cannot see. To do this is to identify a domain in which methodological behaviorism is correct. This result will supplement, not displace, the earlier arguments against behaviorism. That the stategy is appropriate for some explanatory problems does not mean that it cannot be inappropriate for others.

Again, the key idea is *generality*. Let us think, not of explaining token behaviors, but of saying what interesting similarities there may be *among* token behaviors. My simple suggestion is this: *If there are significant commonalities to be discovered among a set of token behaviors – some produced by mentality, other occurring without its help – then mentalism will not be the net in which these common elements can be caught.* If human beings produce a given behavior in virtue of certain cognitive machinery, but members of another species produce that "same" behavior without exploiting such machinery, then we ought to abstract away from the cognitive details peculiar to our species if we want to say what we and this other species have in common. This is not to dany that we have minds, or that our behavior is driven by a cognitive engine. *Methodological* behaviorism involves no such denial. Rather, the point is that we ought not to bring in mental details here. Behavior should be described in ways that are noncommital vis-à-vis intentional etiology; and the general principles that explain the behavior, if they are to be pitched at the requisite level of generality, should themselves not imply mental mechanisms. This pattern of explanation, both in its *explanandum* and in its *explanans*, is just what methodological behaviorism demands.[8]

Evolutionary theory is a context in which this methodological strategy is pursued. Behaviors are described and candidate explanations are suggested in which the question of *proximal implementation* is not addressed. For example, R. A. Fisher (1930) offered an account of why the sex ratio in a population should be 1:1, describing the selective advantage a parent might enjoy in virtue of producing a particular mix of daughters and sons. The basic idea is that when the population's sex ratio is biased towards one sex, a parent will benefit by producing offspring solely of the minority sex. An equilibration process is the result, with a 1:1 sex ratio as the stable equilibrium value.

My point here is not that Fisher's argument is correct,[9] but that it is entirely silent on the mechanisms that individual organisms use to determine the sex ratio of their progeny. In mammals, it is the male who is heterogametic (i.e., males are "XY" while females are "XX"), so one can envision techniques of sex ratio determination in which males produce different ratios of X- and Y-bearing sperm or females produce conditions that differentially affect the viability of the two types of sperm. However, Fisher's argument applies with equal force to the social insects that obey the so-called haplodiploid method of sex determination, according to which all fertilized eggs develop into females, while all unfertilized eggs develop into males. A female stores the sperm she receives from a single fertilization event and then rations them out, thereby determining the sex ratio of her progeny.

Not that evolutionary theory has no interest in explaining the evolution of these various methods of sex determination. My point here is that there is an important style of evolutionary model building that provides insights by abstracting away from the details of proximal mechanism. Fisher's argument purports to tell us something about what a population of mammals and a population of social insects may have in common. A more fine-grained treatment of their special mechanisms of sex determination will not succeed in doing this.

My brief remarks about Fisher's argument do not explicitly join the issue of mentality. However, the pattern of explanation at work in Fisher's argument applies to cases in which one abstracts away from the use of *mental* mechanisms in order to grasp a more abstract generalization.

One line of thinking in evolutionary theory that conforms to this pattern is to be found in sociobiology. Sociobiologists standardly view the mind as a proximate mechanism for producing behaviors that could be brought about in other ways. Natural selection, on this view, favors genes because they code for certain behaviors. It happens that in our species these behaviors are implemented by cognitive structures. Thus, for example, sociobiologists propose to explain incest avoidance by appeal to the genetic advantages of outbreeding. In our species, this behavioral regularity is mediated by the mind; sociobiologists frequently appeal to data about kibbutz children to argue that human beings generally dislike having sexual contact with individuals with whom they grow up. Other species exhibit the same behavior for the same natural selective reasons, but without the intervention of mentality. The sociobiological pattern is as shown below, with "*X*" denoting some nonhuman species of interest:

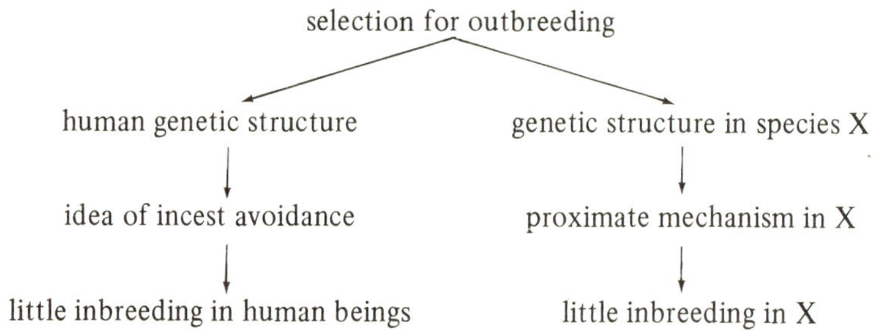

I will not discuss the difficulties that confront this line of argument, on which see Lewontin *et al.* (1984). The pattern, I take it, is familiar enough.

Notice that the sociobiological pattern is a kind of common cause explanation. The idea is that a type of human behavior and a type of behavior in another species trace back to the same kind of cause. Mentality in this picture plays the role of a proximate mechanism. This pattern of explanation may take two forms: It may assert that the two *explananda* trace back to the same *token* event or that they trace back to the same *type* of event. That is, the sameness of characteristics observed in the two species may be a homology or a parallelism. What is ruled out is that the two behaviors have qualitatively different etiologies – that they be merely analogous or convergent.

This common cause form of explanation is standard in sociobiological argument. Territoriality, xenophobia, rape, and slave-making, to name only a few prominent examples, are seen as behaviors that occur among human beings and also in species that presumably do not have a cognitive engine to serve as a proximate mechanism. Critics of sociobiology (e.g., Allen *et al.*, 1976) have been quick to challenge this use of language as anthropomorphic; human behaviors are read into the rest of nature, only to be read back as "natural" and biologically inevitable.

It is important to recognize how this line of criticism should be understood. No evolutionist ought to argue that, since a particular behavior is produced by mentality in human beings, it thereby cannot be found in other species that do not have minds. Mentality *may* be one among numerous possible proximate mechanisms. Whether a behavior can be produced both with and without a mind should be a substantive question, not settled by some general stipulation to the effect that human behaviors are essentially mental. As critics of the critics of sociobiology have rightly remarked, human beings engage in parental care. Indeed, they digest and reproduce. The fact that we use our heads to do these things does not mean that other species that do the same things must follow suit (Wilson, 1976).

Yet, there is a substantive criticism that critics of "anthropomorphic" terminology may have had in mind, one that cannot be so lightly dismissed. Let us define the concept of "slave-making" so broadly that both human beings and the social insects can be said to engage in "the same" behavior. Let us suppose further that the relevant ant species engage in this behavior because there has been natural selection for it. Slave making became prevalent in this species, we are imagining, because slave-makers were thereby more reproductively successful than nonslave-makers, where both these traits are transmitted genetically from parents to offspring. Critics of sociobiology object to the idea that any such explanation accounts for "slave-making" among human beings.

It is a truism that an organism's behaviors, like any phenotypic characteristics, are the joint products of its genetic endowment interacting with its environment. So, when some human organisms engage in slave-making, the behavior is the product of this complex of genes + environment. The genome, we may suppose, has been shaped by the action of evolution, preeminently, perhaps, by natural selection. If selection produces the human genome, and if that genome produces slave-making behavior, does it not follow by transitivity of causality that selection produced the slave-making?

This is not the place to undertake a full-scale consideration of whether causality in this case is transitive (on which see Sober, 1984). Let me just assert now that the inference does not follow. Consistent with the above truisms about genes and environment is the following alternative to the sociobiological story we are considering (Gould, 1977). In the historical period in which the human genome was substantially assembled, there never was selection for slave-making. Rather, there was selection for certain broad-scale cognitive abilities, like the use of language and the ability to reason (what I'll call here, slightly ironically, "general intelligence"). These capacities, once fixed, subsequently encountered numerous different environments in different places and times. In some, slave-making results. But slave-making was an indirect consequence of selection fixing other capacities that then encountered certain specific environments; there was no direct selection for the behavior in question. Rather than the common cause explanation sketched above, critics of sociobiology frequently urge an explanation in terms of separate causation, according to which "slave-making" in ants and human beings are merely analogous. This separate cause explanation is sketched below:

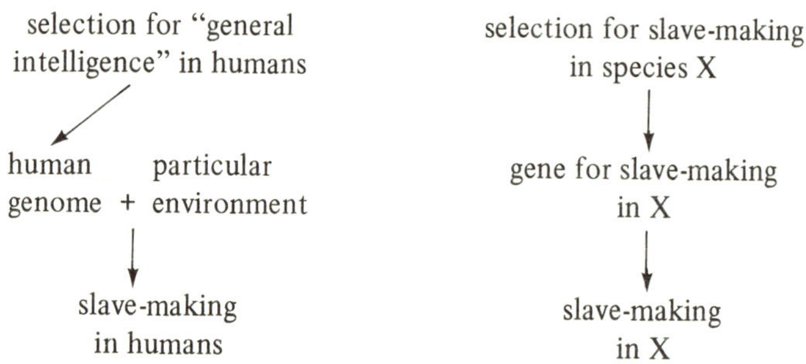

Critics and defenders of sociobiology (including Wilson, 1976, himself) seem to agree that this pattern of explaining slave-making in humans and in ants is vastly more plausible than the common cause form of explanation sketched before.

Notice that this alternative to the sociobiologist's common cause explanation does not reject the idea that human beings and ants engage in the "same" behavior. *Slave-making* is to be explained in both species. The objection, rather, takes the form of an argument to the effect that the behavioral similarities, such as they are, are the product of convergent evolution.[10]

I have sketched two extreme examples. The incest case is perhaps one of the more plausible items in the sociobiological arsenal of common cause explanations. The slave-making example illustrates what an alternative to the common cause structure would look like. The strife over sociobiology to a considerable extent concerns how far the common cause format can be extended – how much of the details of human behavior can be shown to be the upshot of the direct control of natural selection. This question is not answered by the truism that everything we do is the product of the human genome's interacting with the environment, where that genome was shaped by natural selection. For this general remark is equally compatible with the common cause and the separate cause form of explanation.

Take rape as a controversial example. The following passages from Barash (1979) show how sociobiology's common cause explanations are not stopped short by the fact that mentality is a human proximate mechanism (all quoted in Bleier, 1984, pp. 32–33):

> For example, plants with male flowers will "attempt" to achieve as many fertilizations as possible. How is this done? Among other things, they bombard female flowers with incredible amounts of pollen, and some even seem to have specially evolved capacities to rape female flowers, by growing a pollen tube which forces its way to the ovary within each female (p. 30).
>
> Plants that commit rape ... are following evolutionary strategies that maximize their fitness. And, clearly, in neither case do the actors know what they are doing, or why. We human beings like to think we are different. We introspect, we are confident that we know what we are doing, and why. But we may have to open our minds and admit the possibility that our need to maximize our fitness may be whispering somewhere deep within us and that, know it or not, most of the time we are heeding these whisperings (p. 31).
>
> Rape in humans is by no means as simple, influenced as it is by an extremely complex overlay of cultural attitudes. Nevertheless mallard rape and bluebird adultery may have a degree of relevance to human behavior. Perhaps human rapists, in their own criminally misguided way, are doing the best they can to maximize their fitness. If so, they are not that different from the sexually excluded bachelor mallards (p. 55).

I will not examine the objections that can be made against this interpretation of rape (on which see Bleier (1984) and Lewontin *et al.*, Rose, and Kamin (1984). In the present context, I will confine myself to a single point. Even if it were true that a male who now rapes thereby boosts his reproductive success, it would not follow that rape was ever selected for. An alternative theory is that rape is an expression of anger against women and is the product of a mind set that was never shaped in evolutionary time by the reproductive consequences of rape. Rape may be a point in common between human males and mallard ducks no more than slave-making may be a behavior shared by ants and humans.

There is no reason to think that these examples – incest avoidance, slave-making, and rape – must all stand or fall together. The explanations constructed by sociobiologists of these and other examples must be examined one by one on their own merits. Sociobiology is not a single unified theory that issues in a multiplicity of consequences. It is a hodge-podge of similar sounding models that may vary enormously in quality.[11]

It is interesting to note the conflicting roles that these two forms of explanation assign to mentality. In the common cause pattern, the mind does the work of providing a proximate mechanism. It does not count as a causal influence in its own right, but simply implements a behavioral regularity. As we have noted earlier in connection with Fisher's sex ratio argument, it is hardly atypical of evolutionary model building to treat proximate mechanisms in this way.

The alternative form of explanation, on the other hand, sees the presence of mentality as entirely transforming the nature of the evolutionary problem at hand. Not that the mind did not evolve; this truism is not at issue. But a behavior mediated by mentality is thereby connected to a very complex organ; it may no longer be an isolatable unit that natural selection may manipulate on its own. If the respective proximate mechanisms of incest avoidance, or of slave-making, or of rape did nothing else besides trigger these behaviors, then it might be reasonable to view those mechanisms as passive contrivances that have been shaped in evolutionary time simply by the single behaviors they encode. But to see these behaviors as a small sample of the diverse behaviors that can issue from the human mind is potentially to deprive them of their autonomous evolutionary reality. The mind can then no longer be pushed into the background as an incidental proximate mechanism, but rather becomes an evolved mechanism of the first importance.

Before moving on to other styles of explanation, it may be useful to consider how these ideas about the description of behavior in evolutionary theory bear on another example – the concept of altruism. Evolutionists call on organism altruistic when it advances the reproductive interests of others at its own expense.[12] It is a problem of the first importance to see whether altruism is found in nature and, if so, to discover its evolutionary explanation.

What consequences would such accounts have for our understanding of human altruism?

It is true that the vernacular term applies to individuals only when they *intentionally* help others at their own expense. But this fact alone does not remove such human behaviors from evolutionary explanation, even though it does highlight the difference between explaining helping behavior and explaining the intention to help. Intention may, after all, simply function as a proximate mechanism. What would be relevant is determining how the human behaviors affect *reproduction* (this being the evolutionarily relevant senses of helping and hurting) and in seeing whether those behaviors were directly selected for or are off-shoots of more general cognitive structures that themselves evolved for quite different selective (and nonselective) reasons.[13]

Sociobiology attempts to identify the reproductive consequences of various behaviors *and* to read those reproductive consequences back into the history of the evolution of the behavior. Some pitfalls of this enterprise are already quite clear. A trait that now enhances fitness may not have done so in the ancestral past. A trait may enhance fitness for one reason now, but may have done so for a completely different reason ancestrally. However, there is an additional difficulty that sociobiological argument must overcome. To see what this is, let us suppose that incest avoidance is a behavior that augments the reproductive success of incest avoiders and that this is true not just now, but ancestrally. Does it follow that the behavior evolved by natural selection?

The answer is no. Evolution by natural selection requires *genetic* variation as well as *phenotypic* variation. The phenotypic trait in question must be variable in the ancestral population, and, in addition, these phenotypic differences must mirror genetic differences. Without the latter, the enhanced reproductive success of individuals with the trait will not increase the frequency of the trait in the next generation.

It is easy to see how the sociobiological explanation of incest avoidance conforms to this pattern. But to defend this explanation, sociobiologists must discount a quite different explanation. Let us begin with an ancestral population in which all individuals are genetically identical. Because they live in different environments, however, the individuals are phenotypically different. In the present example, this means that some individuals avoid incest while others do not. Suppose further that the trait is *culturally*, rather than genetically, transmitted. Suppose, as is true for many cultural traits (e.g., which language you speak), that cultural transmission leads the trait to run in families (e.g., suppose parents teach children). In this case, the trait in question will increase in frequency, but not because of Darwinian natural selection.

It is, of course, not plausible to think of any ancestral population as genetically homogeneous. But the above cultural line of argument does not require this. Even if there were, somehow, genetic differences between incest avoiders and incest nonavoiders, the argument would still stand, as long as those genetic differences did not cause the behavioral difference in question. For example, suppose that incest avoiders happen to have had a certain blood type in higher frequency, where blood type is genetically determined. In this case, the evolution of the incest taboo will be accompanied by the evolution of blood types. Although both involve a change in frequency of a phenotype, only the latter is "Darwinian." Change in gene frequency mediates the latter, but not the former.[14]

Sociobiology aims to accommodate substantial portions of the human behavioral repertoire in the format of Darwinian evolution. I mention the above alternative, in part, to illustrate one way in which non-Darwinian processes can "mimic" the Darwinian pattern. A trait may influence reproductive success and thereby increase in frequency without the intervention of genes. Notice that if incest avoidance has become prevalent via this non-genetic mode of transmission, there still may be a common element in the stories explaining incest avoidance in human beings and incest avoidance in a species in which the behavior is genetically encoded. The story involved here will again be a setting in which methodological behaviorism is presupposed.

I now want to discuss a third sort of process that is structurally analogous to Darwinian evolution by natural selection. This will provide yet another context in which methodological behaviorism may find an application. Although the two conjectures discussed above about incest avoidance disagree on its mechanism of transmission, they agree that the trait, in our species, confers a fitness advantage.[15] To say that there was selection for the trait here means that the trait caused reproductive success. But traits may increase in frequency in a population without affecting the reproductive success of organisms (or of any other biological units – groups, chromosomes, genes).

Dawkins' (1976) remarks about *memes* suggest how this is possible. Let us think of ideas as infectious agents that can jump from mind to mind in a population. Given the mind-sets found in a society, some ideas will be more infectious than others. They will spread through the population much faster than others. Through time, some ideas will become common while others will go extinct. Not only is it possible for this to happen without any change in gene frequency. In addition, it can happen without the ideas ever making any difference for the reproductive prospects of the organisms in which they occur. Here we have cultural evolution as a process that is entirely autonomous from Darwinian evolution.[16]

There are lines of research now being pursued in evolutionary theory that aim at uncovering what structural similarities there may be between this sort of autonomous cultural evolution and genetically mediated Darwinian evolution. They also conform to the dictates of methodological behaviorism, but do not follow sociobiology in postulating genes for particular mental traits or in thinking that the mental traits of interest have some sort of adaptive significance. The fundamental idea is that the mind can set in motion processes similar to genetic change via Darwinian natural selection. In his book *The Evolution of Cooperation*, Robert Axelrod (1984, pp. 169–170) states the motivation behind this sort of model building quite nicely. Although he calls it "the evolutionary approach," it is important for our purposes to note that it encompasses much more than changes in gene frequencies:

> The evolutionary approach is based on a simple principle: whatever is successful is likely to appear more often in the future. The mechanism can vary. In classical Darwinian evolution, the mechanism is natural selection based upon differential survival and reproduction. In Congress, the mechanism can be an increased chance of reelection for those members who are effective in delivering legislation and services for their constituency. In the business world, the mechanism can be the avoidance of bankruptcy by a profitable company. But the evolutionary mechanism need not be a question of life and death. With intelligent players, a successful strategy can appear more often in th future because other players convert to it. The conversion can be based on more or less blind imitation of the successful players, or it can be based on a more or less informed process of learning.

This avenue of approach does not involve the common cause structure put forward by sociobiologists. The idea, for example, is that changes in gene frequency driven by natural selection and vote trading in Congress have a common structure. It would be altogether gratuitous to claim that senators make deals because they have a gene that causes them to do so, said gene being the product of natural selection. The *etiology* of the mind is not at issue here; what matters are some of the *consequences* it can produce. The explanatory structure is one of analogy, not common cause:

Darwinian natural selection	Rational decision making
↓	↓
genetic structure	behavioral strategy
↓	↓
behavior B	behavior B

Axelrod's (1984) book is an interesting example of how game theoretic explanations can subsume such apparently diverse phenomena within a single format. He describes the kinds of strategies that are optimal for a game that he calls an iterated prisoners' dilemma. Different players pursue their own strategies in the contest. They meet in pairs, interact, and then part ways. The game proceeds for many rounds with players playing against various opponents and sometimes encountering players they have played before. In each encounter, a player must choose between "cooperating" and "defecting." No matter what the other player does, you do better in a given encounter by defecting than by cooperating. However, if both players cooperate, both do better than they would if both defect. The fact that players reencounter each other makes it possible for strategies of cooperation to do better in the long run, even though, in any given encounter, defection is preferable.[17]

Axelrod's argument applies to human beings who use beliefs and desires to assess the moves they might make. So, for example, one of his applications is to be rather surprising level of *non*aggression that apparently obtained during the trench warfare of World War I. But the model also applies to evolving organisms who follow the dictates of their behavioral strategies without the help of minds to tell them what to do. Again the result is to show that the best long-run strategy to follow is one that exhibits a surprising level of cooperation.

Although this game theoretical result has normative import, it also has descriptive and explanatory implications. To be sure, it indicates what one *ought to do*, given certain goals and constraints. But it also describes what a system *will do*, on the hypothesis that it is governed by an optimizing mechanism. The generality of game theory as a descriptive (as well as normative) methodology derives from the fact that optimizing mechanisms can take a variety of forms – human decision making and evolution by natural selection being our two cases in point.[18]

A model that is to apply to both these cases must abstract away from the fact that the trench warfare prisoners' dilemma involves mentality *and* from the fact that the evolutionary case uses natural selection mediated by a Mendelian hereditary mechanism. These are facts specific to the two instances, not something they have in common. Don't be misled by the fact that the players in World War I were the products of natural selection. The point is that they reached their live-and-let-live solution to the iterated prisoners' dilemma game without any change in gene frequencies playing a role.[19]

Game theory is a perfect setting for methodological behaviorism, one that does not involve the controversial assumptions needed by sociobiology's "proximal implementation" picture of mind and culture. The sociobiological explanation of incest avoidance sees that behavior as the product of a specific selection process. But no such story is needed to allow game theory to apply simultaneously to trench warfare and the evolution of cooperation. The behavior exhibited in the trenches, it is assumed, was the product of individual players' desires to stay alive. Although that desire may be thought of as the result of natural selection, game theory requires no such hypothesis. *God* could have made the soldiers one by one; this would not have prevented them from playing an iterated prisoners' dilemma game.

The game-theoretical approach does not assume that the mind is a proximal cause of behaviors that have a more ultimate genetical basis. Sociobiology, at least in its usual formulations, assumes precisely this. The generality of game theory – its ability to subsume games played by individuals with minds and games played by individuals wholly lacking in mentality – arises from the fact that rational deliberation and natural selection are structurally *analogous*. Sociobiology strives for this same level of generality by introducing additional assumptions. Despite these differences, it should be clear that both approaches provide a setting for methodological behaviorism.

4. CONCLUSION

In view of the kinds of criticisms that mentalism raised against behaviorism and that functionalism raised against the identity theory, it is not surprising that methodological behaviorism may have its point. The quest for one sort of generality motivates postulating mental states; the quest for another justifies abstracting away from them. Given our search for various kinds of generalizations that cut across each other, it is to be expected that methodological behaviorism should fail to be *the* technique for explaining behavior. But it is equally clear why it may be *a* technique, and an indispensible one at that. The time has perhaps come for us to think not of behaviorism *versus* mentalism, but of behaviorism *and* mentalism.

ACKNOWLEDGEMENTS

I am grateful to Berent Enc for comments on an earlier draft and to the National Science Foundation and the University of Wisconsin, Madison, Graduate School for financial support.

University of Wisconsin, Madison

NOTES

1 My thanks to Aaron Sober (personal communication) for reminding me of this children's game in which transitivity fails.

2 Claims about general strategies take the form of existence claims ("There exists a stimulus history or a set of beliefs and desires that explains . . ."). Particular explanations conforming to a general strategy are obtained by existential instantiation, as it were. Although claims about general strategy are not "unfalsifiable," they are not crisply refuted by the failure of a single particular explanation. It often takes something like a long run of failure to make it reasonable to cast about for another approach.

3 This, I gather is just what Ryle (1949) does deny. He thinks that behavioral dispositions need not be underwritten by facts about the subject's internal make up. I don't deny that this is "possible" in some fairly denatured sense of possibility. It is likewise "possible" that solipsism is true. If mentalism's only "mistake" were assuming that there must be internal bases for a subject's complex behavioral dispositions, it would be on very firm ground indeed.

4 For the rest of this section, I will take the requirement that the stimulus be described nonmentalistically to mean that it must be described physicalistically. In the next section, we will see that this dichotomy is not inevitable, in that evolutionary models often describe the behaviors they wish to explain in terms of biological properties that are neither mentalistic nor physicalistic in character. The latter point pertains to the so-called "supervenience" of biological properties on physical ones, on which see Sober (1984).

5 For example, ones like Lewis' (1973) that hold that the cause must be necessary for its effect. I find this view implausible for reasons independent of what is at issue here.

6 It is even weaker than a physicalism that claims that each event is *token identical* with some physical event.

7 Putnam (1975) has taken a position on the nature of explanation that does deny this. His idea does for explanation roughly what Lewis' (1973) does for causation. Putnam discusses the case of a peg fitting through a hole in a board. The micro-configuration of the particles in the board and peg, he says, does not explain why the peg fits through the hole, whereas the macro-facts concerning their relative sizes do. I think this claim trades on confusing the question of whether a story explains with whether the story provides an especially clear and simple explanation. It also comes dangerously close to making the mistake of demanding that the *explanans* statements cite only those facts that were necessary for the truth of the *explanandum*.

8 There is a similar point to be made, which I will not develop here, about the explanation of what behaviors have in common when they all issue from cognitive structures, though from different ones. Functionalism in anthropology, for example, might try to describe what an initiation ritual and a religious ceremony have in common, even though the human actors in the two cases have very different motivations. To say that both behaviors "advance group solidarity" and that this explains the fact that such institutions continue to exist may be true, if there is a mechanism that brings the effects of such practices to bear on the persistence of the culture and of those institutions themselves. If this is true, then the uniform functionalist explanation of the two practices abstracts from the mental details peculiar to each.

9 On which see Sober (1984) for discussion of some subsequent work on sex ratio theory.

[10] Furthermore, in keeping with standard practice in phylogenetic reconstruction, biologists will try to redescribe this convergence so that it disappears – i.e., they will look for ways to recode the traits in question so that it is no longer true that human beings and ants are judged to be "the same."
[11] This point is emphasized by Philip Kitcher in his forthcoming book on sociobiology.
[12] See Sober (1984) for a more precise characterization.
[13] Confusing the idea that a trait emerged because of selection and the idea that the trait evolved because there was selection for that very trait has done much to obscure this important question about sociobiological hypotheses. If there was selection for all-purpose problem solving ability and if this trait was correlated with the ability to use abstract moral or mathematical rules, then *selection of* the former characteristic in the course of evolution implies *selection of* the latter; in virtue of the correlation, if individuals with the one trait are, on average, at a reproductive advantage, the same holds true of individuals with the other trait. However, if there was *selection for* the all-purpose ability, this does not imply that there was *selection for* the moral or mathematical ability. For a fuller treatment of the difference between *selection of* and *selection for*, see Sober (1984).
[14] Colwell and King (1983) provide a perspicuous development of this idea in their discussion of "the sower's fallacy."
[15] I should point out that parent-offspring and brother-sister mating in common in many species. A careful explanation, sociobiological or otherwise, of incest avoidance should not begin with the blanket assumption that such mating patterns are always deleterious. In addition, it is important to recognize that it is one thing to explain why a species is outbreeding, quite another to explain why it achieves this behavior by the proximate mechanism of a social taboo.
[16] Although Cavalli-Sforza and Feldman (1981) are mainly interested in the evolution of traits affecting the reproductive success of organisms that can be both genetically and culturally transmitted, they emphasize (p. 49) that epidemiological models apply to "epidemics and rumors" with equal force.
[17] The connection of the prisoner's dilemma with frequency dependent selection and group selection is discussed in Sober (1984).
[18] See Maynard Smith (1978) for a discussion of optimality arguments in evolutionary theory and Sober (1984) for a discussion of the sense in which natural selection may be thought of as an optimizing process.
[19] Note that the nonaggressive behavior of combatants could have been produced either by having noncooperative soidiers die, or by having individuals modify their behavior via deliberation and learning. Either a "variational" or a "developmental" process (to use Richard Lewontin's terminology) would have sufficed. See Sober (1984) for discussion.

REFERENCES

Allen, E. *et al.*: 1976, 'Sociobiology – Another Biological Determinism', *BioScience* **26**. Reprinted in A. Caplan (ed.), *The Sociobiology Debate*, Harper & Row, New York, 1978, pp. 280–290.
Axelrod, R.: 1984, *The Evolution of Cooperation*, Basic Books, New York.
Barash, D.: 1979, *The Whisperings Within*, Harper & Row, New York.

Bleier, R.: 1984, *Science and Gender*, Pergamon Press, New York.

Cavalli-Sforza, L. and Feldman, M.: 1981, *Cultural Transmission and Evolution: A Quantitative Approach*, Princeton University Press, Princeton.

Colwell, R. and King, M.: 1983, 'Disentangling Genetic and Cultural Influences on Human Behavior: Problems and Prospects', in D. W. Rajecki (ed.), *Comparing Behavior: Studying Man Studying Animals*, Lawrence Erlbaum, Hillsdale, N.J., pp. 227–250.

Dawkins, R.: 1976, *The Selfish Gene*, Oxford University Press, Oxford.

Dennett, D.: 1978, 'Skinner Skinned', in *Brainstorm: Philosophical Essays on Mind and Psychology*, Bradford/MIT Press, Cambridge, pp. 53–70.

Fisher, R. A.: 1930, *The Genetical Theory of Natural Selection*, Dover, New York, 2nd edition, 1957.

Fodor, J.: 1968, *Psychological Explanation*, Random House, New York.

Fodor, J.: 1975, *The Language of Thought*, Thomas Crowell, New York.

Gould, S.: 1977, 'Biological Potential vs. Biological Determinism', in *Ever Since Darwin*, W. W. Norton, New York, pp. 250–259.

Lewis, D.: 1973, 'Causation', *Journal of Philosophy* 70, 556–67.

Lewontin, R., Rose, S., and Kamin, L.: 1984, *Not In Our Genes: Biology, Ideology, and Human Nature*, Pantheon, New York.

Maynard Smith, J.: 1978, 'Optimization Theory in Evolution', *Annual Review of Ecology and Systematics* 9, 31–56. Reprinted in E. Sober (ed.), *Conceptual Issues in Evolutionary Biology*, Bradford/MIT Press, Cambridge, 1984, pp. 289–316.

Place, U. T.: 1962, 'Is Consciousness a Brain-process?', in V. C. Chappell (ed.), *The Philosophy of Mind*, Prentice-Hall, Englewood Cliffs.

Putnam, H.: 1967, 'The Nature of Mental States', in *Mind, Language, and Reality*, Cambridge University Press, Cambridge, 1975, pp. 139–152.

Putnam, H.: 1975, 'Reductionism and the Nature of Psychology', in *Mind, Language, and Reality*, Cambridge University Press, Cambridge.

Ryle, G.: 1949, *The Concept of Mind*, Hutchinson, London.

Smart, J. J. C.: 1959, 'Sensations and Brain Processes', *Philosophical Review* 68, 141–156.

Sober, E.: 1983, 'Mentalism and Behaviorism in Comparative Psychology', in D. W. Rajecki (ed.), *Comparing Behavior: Studying Man Studying Animals*, Lawrence Erlbaum, Hillsdale, N.J., pp. 113–142.

Sober, E.: 1984, *The Nature of Selection: A Philosophical Inquiry*, Bradford/MIT Press.

Tinbergen, N.: 1951, *The Study of Instinct*, Clarendon Press, Oxford.

Wilson, E.: 1976, 'Academic Vigilantism and the Political Significance of Sociobiology', *Bioscience* **26**, 183–190. Reprinted in A. Caplan (ed.), *The Sociobiology Debate*, Harper & Row, New York, 1978, pp. 291–303.

PAUL THOMPSON

SOCIOBIOLOGICAL EXPLANATION AND THE TESTABILITY OF SOCIOBIOLOGICAL THEORY

ABSTRACT. During the last decade there has been a lively debate about the degree to which sociobiological theory is untestable due to the highly speculative and *ad hoc* nature of sociobiological explanations and, usually by inference from the explanatory cases, predictions – especially those concerning human behavior. In this paper, I examine this problem within the framework of the semantic conception of theories. I shall argue that a semantic conception of theories provides a more thoroughgoing analysis of this problem than the more familiar syntactical conception of theories. In particular, a semantic conception more accurately represents the relation between a theory and phenomena which, in this case, results in a clearer understanding of: the nature of the problem of the testability of sociobiological theory, the difference in explanatory success of the theory when applied to insects and to humans, and the kind of additional knowledge required in order to apply the theory successfully to human behavior.

Although speculation about the possibility of providing evolutionary explanations of social behavior go back as far as Darwin, sociobiology as a discipline is extremely young – coming of age with the publication of E. O. Wilson's book *Sociobiology: The New Synthesis*[1] in 1975. During this brief history there has been a lively debate over its status as a scientific theory[2]. An important issue in the debate, and the one that I shall examine in this paper, is the degree to which sociobiological theory is untestable (hence, unscientific) due to the highly speculative and *ad hoc* nature of its explanations and, usually by influence from explanatory cases, predictions – especially those concerning human behavior.

I shall explicate and examine certain aspects of the ways in which sociobiological theory is applied to behavior, for the purpose of explaining behavior[3], within the framework of a relatively new account of the structure of scientific theories called the semantic view (or conception) of theories. I shall argue that among the numerous virtues of this account, as compared with the more often employed syntactical account, is the more accurate understanding that it provides of the relationship between a theory and phenomena.

I shall then relate this general virtue to the above controversy by arguing that the alleged lack of testability of sociobiological theory based on the highly speculative and *ad hoc* nature of its explanations can, on a semantic

James H. Fetzer (ed.), Sociobiology and Epistemology, 201–215.

account of theories, be seen to be the result of problems with one or more of the theories employed in relating sociobiological theory to the phenomenal world and not with sociobiological theory itself. That is, the highly speculative and *ad hoc* nature of sociobiological explanations is a function of the unavailability of certain theories required in the application of sociobiological theory to behavior – especially human behavior. Hence, this feature of sociobiological explanations of human behavior provides no grounds for rejecting sociobiological theory, because any lack of testability is not a function of the character of sociobiological theory but rather of the character of certain explanatory applications of it.

I hasten to add, however, that this problem does provide grounds for exercising extreme caution – considerably more than has been exercised to date – since, as a host of writers have pointed out[4], sociobiological theorizing and explaining takes place within a social and political context and, as has already happened[5], the explanations of sociobiologists will be used to support ideological and political positions.

I

In various forms the semantic conception of theory structure has been developed by Patrick Suppes[6], Bas van Fraassen[7], Frederick Suppe[8], Joseph Sneed[9] and Wolfgang Stegmuller[10]. And recently it has been applied to aspects of evolutionary biology by John Beatty[11], Elizabeth Lloyd[12] and myself[13]. In what follows, I shall, in the few cases where a particular approach within the semantic view needs to be assumed, couch my remarks in terms of the state space approach of van Fraassen and Suppe.

On semantic view accounts, a theory is an extralinguistic entity which describes physical systems by specifying the possible states and behaviors of systems. A physical system is related to the phenomenal world by asserting that the possible states and behaviors of the empirical systems within the intended scope of the theory are isomorphic to those of a particular physical system specified by the theory. On the state space approach of van Fraassen and Suppe, the physical system is characterized in terms of: a definition of a state space (an *n*-dimensional Cartesian space), laws defining the possible sets of states in the state space (laws of coexistence), laws defining the physically possible trajectories in the state space (laws of succession[14]), and laws defining the physically possible results of interactions of the systems defined by the theory with other systems.

On the semantic view, theories do not describe the behavior of phenomena but of physical systems. Hence, laws do not directly describe the behavior of phenomena but describe the behavior of physical systems. Physical systems are idealized systems in that they are abstractions from phenomena which characterize the ways in which phenomena would behave if the idealized conditions of the system were met[15].

There are several differences between this account of the structure of scientific theories and the syntactical account of the logical empiricists, dubbed by Putnam[16] as the received view, on which theories are axiomatic deductive structures, the theoretical terms of which are partially interpreted by means of correspondence rules[17]. The difference most relevant to the issues discussed in this paper is the more accurate account provided by the semantic view of the way in which a theory is applied to phenomena[18].

On a received view account, the ways in which theories are applied to phenomena are specified by the correspondence rules which partially interpret the theoretical terms of the calculus. These correspondence rules are part of the theory. Hence, unlike the semantic view, the procedures for applying the theory to phenomena are part of the theory itself. This aspect of the received view obscures many important epistemological features of the relationship between a theory and phenomena. First, it misleadingly entails that any changes in the procedures for applying the theory will result in a new theory and not, as seems a more accurate description, in new ways of applying the same old theory.

Second, specifying the procedures for applying the theory to phenomena by means of correspondence rules which are part of the theory does not provide an accurate representation of the kinds of ways in which theories are applied to phenomena. For example, as Kenneth Schaffner[19] has convincingly argued, it does not provide an accurate account of the ways in which several theories can be used to provide a causal sequence in which borrowed theories link a designated theory with observation reports. The correspondence rules of the borrowed theories, when used to provide a causal sequence linking a designated theory with phenomena, are different in kind from the correspondence rules of the designated theory. For one thing, although they are used to relate the designated theory to phenomena, they are not part of the designated theory.

Third, the received view account provides a distorted picture of the experimental procedures employed in relating theories to phenomena. As Patrick Suppes[20] has argued, there is a hierarchy of theories which mediate between a designated theory and an experimental situation.

Suppes has identified three theories which, along with numerous unstated *ceteris paribus* conditions, mediate between a physical theory (which stands at the top of the hierarchy of theories) and phenomena: the theory of the experiment, the theory of data and the theory of experimental design (which is where the *ceteris paribus* conditions are relevant). Possible realizations of the theories are models of the theories.

The theory of experimental design is employed to structure an experiment. Randomization of the assignment of subjects to groups, materials to be used, methods of administering drugs or tests, the relationship of the output of instruments to the characteristics and behavior of phenomena, etc.) are part of the theory of experimental design. In structuring the experiment, the theory of experimental design will frequently employ other theories in causal sequences of the kind outlined by Schaffner. For example, relating the behavior of chromosomes to observed images produced by a light microscope under certain experimental conditions will require employment of a theory of optics as well as a theory about staining tissues, etc. Part of what the theory of experimental design does is to specify some of the causal sequences which Schaffner has identified as being involved in relating theories to phenomena. In addition to the stated structure produced by the theory of experidmental design, numerous unstated *ceteris paribus* conditions (e.g., noise levels, the amount of sleep a subject had the previous night, background electromagnet radiation, etc.) will be taken into account at the level of implementing the conditions described by a realization of the theory of experimental design.

The theory of data is employed to relate the relevant entities and their behavior as a phenomenal system to the relevant parameters of a particular physical system which is claimed to be isomorphic to the phenomenal system. That is, a realization of the theory of data will identify entities and their behavior at the phenomenal level with the parameters of a particular physical system and will describe conditions of adequacy of fit (usually statistical tests) between the phenomena and the parameters of the physical system. As with theories of experimental design, theories of data will often employ other theories in causal sequences of the kind described by Schaffner. A realization of the theory of data enables a comparison of the results of the experiment, which are in terms of observed entities and their behavior, with the predictions of the physical system described by the theory of the experiment. This comparison is usually carried out in terms of statistical tests of goodness of fit.

The theory of the experiment appears, for Suppes, to be employed to select from all the possible physical systems described by the theory those which, if the theory is true, represent the particular experimental framework under investigation. That is, it is used to induce a class of causally possible physical systems which represent the experimental framework[21].

The semantic view accommodates these complexities of applying theories to phenomena because it distinguishes between theories (which are extra-linguistic devices for specifying a class of physical systems) and the application of theories. That is, borrowed theories used in applying the theory remain independent of the applied theory. The received view fails to accommodate them without distortion because the ways of applying theories (which are linguistic entities from which statements about the behavior of phenomena can be deduced) are part of the theory via the correspondence rules. Hence, borrowed theories cannot, in any natural way, be employed in applying a theory. With this background, I turn now to the controversy over the status of sociobiological theory.

II

Since the publication of *Sociobiology: The New Synthesis* numerous critics have raised a wide range of objections to sociobiology[22]. Some of these objections involve important methodological criticisms of it[23]. And, some of these methodological criticisms are used as a basis for arguing that sociobiological theory is untestable and, therefore, fails to be a genuine scientific theory. The most common methodological criticism used for this purpose is that sociobiological explanations (and, by inference, predictions) are highly speculative and *ad hoc*. A central reason for this, according to the critics, is that the genetic basis of the phenomena which are being explained in terms of sociobiological theory is assumed without evidence and any evidence that might suggest non-genetic transmission (as for example, the incredible short time span – in evolutionary/genetic terms – during which cultural change takes place) is rendered impotent by the postulation of *ad hoc* mechanisms (for example, the multiplier effect). These features, they contend, make sociobiological theory immune to evidence (hence, unscientific) and sociobiological explanations little more than fascinating storytelling. With slight differences of emphasis, Richard Burian, Stephen J. Gould and Richard C. Lewontin[24] have developed this criticism.

I shall argue that whole this is a methodological problem for sociobiology, on a semantic view account of theories, it can be clearly identified as a problem with the application of sociobiological theory to phenomena (i.e., behavior, specifically human behavior) and not with the theory itself.

Specifically, it is a problem with the causal sequence of theories employed in relating sociobiological theory to human behavior. This analysis, I contend, is a more thoroughgoing analysis than that provided by the critics and than is possible on a received view account of theories. One important outcome of this anlaysis will be a plausible account of why sociobiological theory is so successfully employed in explaining animal behavior, especially insect behavior – a success freely admitted by some of the critics[25] – and yet so apparently problematic when employed to explain human behavior (and perhaps that of some other higher organisms).

My strategy in explicating the problem involved in applying sociobiological theory to human behavior is two staged. First, I examine relevant logical and epistemological aspects of the successful application of evolutionary theory to the phenomenon of sickle-cell anemia – an application which results in a widely accepted explanation of the persistence of sickle-cell anemia. Second, I examine, in light of the discussion of the sickle-cell anemia case, a controversial application of sociobiological theory to human behavior, namely, that of explaining the persistence of homosexual behavior in human populations.

Consider the case of a person who manifests impaired growth and development, increased susceptibility to infection, severe abdominal pain with normal bowel sounds and no rebound tenderness, paleness and physical weakness. These symptoms are explained in certain cases by reference to hemolytic anemia resulting from the sickling of red blood cells. This explanation appeals to a theory of physiology which describes, among other things, the structure, internal behavior and function of red blood cells in the body as well as the physiological effects of an abnormal red blood cell structure. The sickling of the red blood cells is explained by reference to the fact that, in the case of those with sickle-cell anemia, the nucleotide sequence, which determines the structure of the protein 'hemoglobin', codes for the amino acid 'valine' in the sixth position on the β-chain instead of 'glutamic acid' as is normally the case. This substitution, under reduced oxygen tension, results in the valines at positions 6 and 1 forming a hydrophobic association which leads to a conformation that stacks in such a way as to distort the erythrocyte and thus cause the sickle shape. This explanation appeals to molecular biological theory which describes, among numerous other things, how the nucleotide sequences of DNA determine the structure of proteins and to biochemical theory which describes the behavior of biochemical structures. Hence, by appeal to a number of different theoretical frameworks, a connection between the physical characteristics cited and a sequence of nucleotides is made and the physical characteristics are explained.

Given the character of the physical condition and the high mortality rate associated with it, an explanation of the persistence of this physical condition has been sought. The current explanation of this persistence refers to the increased fitness, in malaria infested environments, of individuals who are heterozygous for the sickle-cell allele (i.e., have one allele for sickle-cell hemoglobin and one for normal hemoglobin at the same locus). That is, the heterozygote has slightly increased protection against falciparum malaria and hence, in malaria infested areas, gains a selective advantage over either homozygote – one will suffer from sickle-cell anemia and the other will have less protection against malaria. This explanation involves an application of evolutionary theory to the phenomenon of sickle-cell anemia. Evolutionary theory describes the laws of succession governing the behavior of alleles, the effects of selection on the genome, the laws of succession under various modes of selection at various levels, etc. And it is in terms of the parameters (alleles, populations, etc.), and the laws governing the behavior of a physical system described by the theory that the explanation is given.

The application of evolutionary theory to sickle-cell anemia in this explanation is mediated by the causal sequence of theories outlined in the above discussion of the explanation of the physical condition. That is, the physical condition described above is related to the physiological condition of sickled red blood cells by a theory of physiology. The physiological condition is, in turn, related to the genotype of the organisms involved by molecular genetical theory and biochemical theory. What in effect this causal sequence of theories does is to causally relate the physical condition to the genotype. This causal relationship is an important feature in the application of evolutionary theory to the sickle-cell anemia case in the following way.

The behavior of the physical systems described by evolutionary theory is a function of laws of succession governing the behavior of genotypes under various conditions, including selection, in a state space. In the case of sickle-cell anemia, among the parameters and laws which describe the behavior of the relevant physical system are gene loci, alleles, selection coefficients, and the Hardy-Weinberg law of succession. In order to relate the behavior of this physical system to the behavior of organisms at the phenomenal level, the above outlined causal sequence of theories must be employed to relate the parameters of the physical system (i.e., alleles, etc.) and their behavior to those of the physical conditions of organisms (i.e., pain, bowel sounds, physical weakness, susceptibility to infection, resistance to malaria, etc.) which are, in turn, related to reproductive success.

Consider now the case of homosexual behavior and the sociobiological explanation of its persistence in the population despite the presumed evolutionary disadvantage of the behavior – it is assumed that those who engage in homosexual behavior will leave few, if any, offspring and, therefore, like those with sickle-cell anemia, are less fit, from an evolutionary point of view, than those who engage in heterosexual behavior. There are several sociobiological explanations of this phenomenon. The one that I shall discuss is similar in structure to the above explanation of the persistence of sickle-cell anemia. That is, persons heterozygous with respect to a homosexuality allele are more fit than either homozygote (they might, for example, have a higher libido and, hence, leave more offspring than either homozygote). Hence, homosexuality persists in populations because in heterozygotes the allele does not result in homosexual behavior and it confers on its bearer a selective advantage over homozygotes for non-homosexual behavior.

There are, as the critics have pointed out, several problems with this explanation[26]. The problem relevant to this paper concerns the ungrounded assumption of a genetic basis to homosexual behavior. This problem is an instance of a more general problem, namely, that underlying sociobiological explanations of behavior is the *assumption* that there is a genetic basis to the behavior being explained. And, in the case of human behavior, this assumption frequently is ungrounded, as it is with this sociobiological explanation of homosexual behavior. Its lack of grounding, however, is not, as the critics have claimed, a result of a lack of evidence but is a result of the absence of other theories required for the theory's application to human behavior. The homosexuality case illustrates this as follows.

The sociobiological explanation of the persistence of homosexual behavior (which is taken to be deleterious from an evolutionary point of view) involves an application of sociobiological theory to homosexual behavior by means of an assumed causal sequence of other theories. This causal sequence of theories is quite similar to the one employed in applying evolutionary theory to sickle-cell anemia and the explanation of its persistence. Also, it is required in this case for the same reasons as the causal sequence of theories was required in the sickle-cell case.

The important difference between the two cases is that a promissory note instead of the required theories is offered in the homosexual case, whereas the required theories are available in the sickle-cell case. That is, there is no physiological theory that causally relates physiological states and behaviors to homosexual behavior and, in the absence of such a theory, the genotype cannot be related to the required physiological states and their behavior because the relevant states cannot be identified. How serious is this problem?

One might be tempted to point out that the sickle-cell anemia case is unusual. It is rare that such a complete causal account of the physiological states can be given in terms of molecular genetical states by reference to molecular genetical and biochemical theory. In addition, it is seldom the case that more than a sketch of a causal account of a physical condition can be given in terms of physiological states and the behavior of physiological systems. Hence, it might appear that the demands being made on sociobiological explanation are overly stringent. There is, however, a good reason why this is not so.

Although it is seldom the case that complete causal accounts can be given for any of the stages in the causal sequences employed in applying a theory to phenomena, in most cases there is at least a widely accepted theoretical framework which grounds the sketchy causal accounts, provides direction to attempts to flesh out the accounts and provides a basis for assessing the relevance of experimental data to the account.

Consider, for example, an application of evolutionary theory to anatomical phenomena in order to explain anatomical differences or similarities among populations or species, or to explain a phylogenetic pattern of anatomical development. This application will involve a causal sequence of theories among which will be molecular genetical theory and embryological theory (or theories). Nothing even approaching a complete causal account of the relation between adult anatomy and the zygote is possible, but the theoretical framework(s) provided by embryology grounds the sketchy accounts that can be given, provides direction to attempts to expand the accounts and provides a framework for assessing the relevance of experimental data. Other cases (such as the molecular genetical one) will be similar.

In the case of human behavior, however, we do not have a widely accepted theoretical framework relating human behaviors, like homosexuality, to physiological states and their behavior. If we did, molecular genetical theory would then serve the above mentioned functions in causally relating the physiological states and their behavior to molecular genetical states and their behavior. While certain human behaviors have been related, by current physiological theory, to physiological states (e.g., wake/sleep patterns[27]), the vast array of social and individual human behaviors cannot be related even in a sketchy way to physiological states by reference to current physiological theory. Indeed, unlike most human behavior, sleeping and waking are manifestations of neural activities in the way that sickle-cell anemia is a manifestation of oxygen deprivation to cells. Current physiological theory describes physical systems in which these outcomes are a function of physiological states and behaviors. Current physiological theory does not describe physical systems in which homosexuality, writing articles, etc. are a function of physiological states and behaviors (though, as I indicate below, some cases of aggressive behavior, along with a limited number of other behaviors may be able, on current physiological theory, to be related to physiological states). Perhaps in the future current theories of physiology will be expanded, revised or replaced such that an adequate theory will be available. At present no such theory exists. It is this problem, not the sketchiness of the accounts, that makes the application of sociobiological theory to human behavior methodologically suspect.

In the case of insect behavior and a considerable amount of animal behavior, however, current physiological theory does describe physical systems which have these behaviors as outcomes of physiological states and behaviors. This is because the behavior of animals discussed by sociobiologists is quite clearly the outcome of physiological states and behaviors of the endocrine system, neurological system or some other physiological system. Consider, for example, Wilson's discussion of hormones and aggression[28]:

> The endorcine system of vertebrates acts as a relatively coarse tuning device for the adjustment of aggressive behavior. The interaction of the several hormones in this control are complicated. ... However, they can be understood readily if the entire system is viewed as comprising three levels of controls: the first determines the state of preparedness (androgen, estrogen and luteinizing hormone). ... The level of preparedness to fight is what we usually refer to as aggressiveness, in order to contrast it with the act of aggression. Aggressiveness, as Rothballer (1967) has said, is a threshold. It can be measured either by the amount of the provoking stimulus required to elicit the act or by the intensity and prolongation of the act in the face of a given stimulus. The class of hormones most consistently associated in the vertebrates with heightened aggression is the androgens, which are 19-carbon steroids, with methyl groups at C-10 and C13, secreted by the Leydig cells of the testes.

Aggression on this account is like withdrawing one's hand when it is pricked by a sharp object. There is a stimulus which elicits a neurological process which results in muscle contraction which results in the behavior of withdrawing the hand. In the case of aggression, there is a stimulus which elicits an endocrine process already primed to go under such stimuli, which results in physiological changes which result in aggressive behavior.

No doubt some aspects of human aggression can be placed within this framework since, like other vertebrates, we have the same hormonal processes as involved in animal aggression. Most human behaviors, however, are not readily able to be fitted into such physiological accounts, whereas insect behavior and a considerable amount of animal behavior can be so fitted. Even in the case of aggression, human aggression is considered to be substantially more complex, involving factors not amenable to physiological explanation. As indicated above the central difference between humans and insects appears to be that human behavior involves features that are not capable of being related to physiological states. It may be that this difference will, in the future, be shown illusory. For now, however, it remains an important difference between the two cases – a difference which affects the application of the theory to phenomena. Having now outline the character of the problem let me draw the threads together in some concluding remarks.

III

There are several results which emerge from this paper. First, the discussion locates the problems outlined by the critics of sociobiology as problems with the application of the theory to humans and not, as the critics have claimed, with the theory itself. This makes intelligible the fact that the theory seems to explain very successfully a wide range of animal behaviors but is much less successful, if at all successful, in explaining human behavior. If the problem is with the theory, as the critics claim, then it is difficult to understand just why the theory is so successful in non-human contexts. If, however, the problem is with the application of the theory to phenomena such that, in the case of humans, other theories required in order to apply sociobiological theory to human behavior are not available whereas, in the case of animals, they very frequently are available, the differences in explanatory success are easily accounted for.

Second, the discussion, by making clear the nature of at least one problem with human sociobiology, makes clear that what is needed are theories that relate human behavior to physiology. And this makes clear the limitations of human sociobiology without denying the scientific status of the theory and without denying that some human behaviors may be explanable by reference to sociobiological theory. That is, in those case where theories which relate human behavior to genetically based physiological processes exist or become available, sociobiological explanation will be possible. In those cases where no such theory exists or where other theories exist which relate the behavior to physiological processes which are related by other theories to environmental (broadly understood) factors and not to genes, then no sociobiological explanation will be possible.

The general point about the need for a genetic basis to be established is, consequently, placed in a different light by making clear that this is not a defect of the theory but of its application and making clear that what is needed is not simply more evidence but other theories which are quite independent of sociobiological theory. The success in applying sociobiological theory to insects indicates the scientific value of the theory and constitutes the arena of testability. The inability to apply it to much of human behavior indicates the limitations of its application. It does not, however, indicate that the theory is untestable and unscientific but only that human behavior is not yet a fruitful arena of testing and, at present, cannot be adequately explained by the theory.

Third, the above discussion, by drawing out some of the complexities involved in sociobiological explanations of human behavior, has illustrated the richness of semantic accounts of theory structure and the more thorough-going analysis of explanation and testability that they make possible.

University of Toronto

NOTES

[1] E. O. Wilson, *Sociobiology: The New Synthesis*, Belknap Press, Cambridge, 1975.

[2] See E. Allen *et al.*, 'Sociobiology: Another Biological Determinism', *Bioscience* **26** (1976), 182–186; R. M. Burian, 'A Methodological Critique of Sociobiology', in A. L. Caplan (ed.), *The Sociobiology Debate*, Harper & Row, New York, 1978, pp. 376–395; S. J. Gould, 'Sociobiology: The Art of Storytelling', *New Scientist* **80** (Nov. 16, 1978), 530–533; R. Lewontin, 'Sociobiology: A Caricature of Darwinism', in F. Suppe and P. D. Asquith (eds.), *PSA 1976*, vol. 2, Philosophy of Science Association, East Lansing, 1977, pp. 22–31; M. Ruse, 'Sociobiology: Sound Science or Muddled Metaphysics?' in F. Suppe and P. D. Asquith (eds.), *PSA 1976*, vol. 2, Philosophy of Science Association, East Lansing, 1977, pp. 48–73; M. Ruse, *Sociobiology: Sense or Nonsense*? D. Reidel, Dordrecht, 1979; P. Thompson, 'Is Sociobiology A Pseudoscience?' in P. D. Asquith and R. N. Giere (eds.), *PSA 1980*, vol. 1, Philosophy of Science Association, East Lansing, 1980, pp. 363–370.

[3] The comments and arguments in this paper concentrate on explanation with little to no reference to prediction which is the concept most often associated with testability. I have concentrated on explanation for two reasons. First, those who have raised this criticism of sociobiology (see, for example, *op. cit.* 'Sociobiology: A Caricature of Darwinism' and *op. cit.* 'Sociobiology: The Art of Storytelling') have expounded it mostly in terms of the highly speculative and *ad hoc* character of sociobiological explanation. Second, nearly all the comments and arguments in this paper relating to explanation can be straightforwardly applied to prediction. Prediction and explanation are, with regard to the features discussed in this paper, symmetrical.

[4] See, for example, S. J. Gould, 'Biological Potential *vs.* Biological Determinism',

Natural History Magazine (May 1976); *op. cit.* 'Sociobiology: The Art of Storytelling'; P. Thagard, 'Critical Notice of Michael Ruse, *Sociobiology: Sense or Nonsense*?' *Canadian Journal of Philosophy* **11** (1981), 751–759.

5 See the references in note 4 for examples.

6 P. Suppes, *Introduction to Logic*, Van Nostrand, Princeton, 1957; and P. Suppes, 'What is a Scientific Theory?' in S. Morgenbesser (ed.), *Philosophy of Science Today*, Basic Books, New York, 1967, pp. 55–67.

7 B. C. van Fraassen, 'On the Extension of Beth's Semantics of Physical Theories', *Philosophy of Science* **37** (1970), 325–339; B. C. van Fraassen, 'A Formal Approach to the Philosophy of Science', in R. E. Colodney (ed.), *Paradigms and Paradoxes*, University of Pittsburg Press, Pittsburg, 1974, pp. 303–366; and B. C. van Fraassen, *The Scientific Image*, Oxford University Press, New York, 1980.

8 F. Suppe, *The Meaning and Use of Models in Mathematics and the Exact Sciences: A Study in the Structure of Exact Scientific Theories*, University Microfilms International, Ann Arbor, 1967 (Ph.D. Dissertation); F. Suppe, 'What's Wrong with the Received View on the Structure of Scientific Theories?', *Philosophy of Science* **39** (1982), 1–19; F. Suppe, *The Structure of Scientific Theories*, 2nd edn., University of Illinois Press, Chicago, 1977; F. Suppe, 'Theoretical Laws', in M. Przelecki, K. Szaniawski, and R. Wojcicki (eds.), *Formal Methods in the Methodology of Empirical Science*, Ossolineum, Wroclaw, 1976, pp. 247–267.

9 J. Sneed, *The Logical Structure of Mathematical Physics*, D. Reidel, Dordrecht, 1971.

10 W. Stegmuller, *The Structure and Dynamics of Theories* Springer, New York, 1976; and W. Stegmuller, *The Structuralist View of Theories* Springer, Berlin, 1976.

11 J. Beatty, 'Optimal-Design Models and the Strategy of Model Building Evolutionary Biology', *Philosophy of Science* **47** (1980), 532–561; and J. Beatty, 'What's Wrong with the Received View of Evolutionary Theory?' in P. D. Asquith and R. N. Giere (eds.), *PSA 1980* vol. 2, Philosophy of Science Association, East Lansing, 1981, pp. 397–426.

12 E. A. Lloyd, 'A Semantic Approach to the Structure of Population Genetics', *Philosophy of Science* **51** (1984), 242–264.

13 P. Thompson, 'The Structure of Evolutionary Theory: A Semantic Approach', *Studies in History and Philosophy of Science* **14** (1983), 215–229.

14 Suppe has also introduced the concept of a law of quasi-succession *op. cit.* 'Theoretical Laws'.

15 For a more detailed account of the nature of physical systems on a state space approach see, *op. cit.*, 'What's Wrong with the Received View of Scientific Theories?'; *op. cit.*, 'The Structure of Evolutionary Theory: A Semantic Approach'; or *op. cit.*, 'On the Extension of Beth's Semantics of Physical Theories'.

16 H. Putnam, 'What Theories Are Not', in E. Nagel, P. Suppes and A. Tarski (eds.), *Logic, Methodology and Philosophy of Science*, Stanford University Press, Standford, 1962, pp. 240–251.

17 Almost all of the previously cited articles which deal with the structure of theories discuss one or more of these differences. The best general account is found in *op. cit., The Structure of Scientific Theories.*

18 John Worrall (J. Worrall, 'Review Article: An Unreal Image', *British Journal for the Philosophy of Science* **35** (1984), 65–80.) has correctly pointed out that there is no genuine logical incompatibility between syntactical and semantic accounts of theory structure. There are, however, as many writers have argued, significant differences in

the ways in which the two accounts represent the role of theories in scientific practice. In large measure these differences result from the fact that semantic accounts, unlike received view accounts, do not characterize theories in terms of linguistic structures but rather in terms of the behavior of systems. Hence, the ways in which theories change, the ways in which they explain and predict phenomena and the ways in which they are confirmed are understood differently on the two accounts. Worrall is, no doubt, correct when he points out that each account has its respective advantages but, as many of us who have developed and applied semantic accounts have argued, the received view account has serious disadvantages which are not shared by semantic accounts. Most of these disadvantages have to do with the way in which the received view account simplifies, misrepresents and distorts the actual practice of scientists in forming, employing and confirming theories.

[19] K. F. Schaffner, 'Correspondence Rules', *Philosophy of Science* **36** (1969), 280–290.

[20] P. Suppes, 'Models of Data', in E. Nagel, P. Suppes and A. Tarski (eds.), *Logic Methodology and Philosophy of Science: Proceedings of the 1960 International Congress*, Stanford University Press, Stanford, 1962, pp. 252–261; and *op. cit.*, 'What is a Scientific Theory?'.

[21] Suppe (F. Suppe, 'Theories and Phenomena', in W. Leinfellner and E. Kohler (eds.), *Developments in the Methodology of Social Science*, D. Reidel, Dordrecht, 1974, pp. 45–91.) considers it the function of the theory itself to determine the class of causally possible physical systems ("Indeed, one function of a scientific theory is to determine what the class of causally possible physical systems is." (p. 48)). Hence, the function of Suppes' theory of the experiment is, for Suppe, in this particular paper, performed by the theory which is being applied to phenomena. Suppe employs the term 'theory of the experiment' to describe the ensemble of a theory of data, a theory of experimental design and *ceteris parabus* clauses used to relate a theory to phenomena (see pp. 71–80 and note 41, p. 90).

[22] See, for example, M. D. Sahlins, *The Use and Abuse of Biology*, University of Michigan Press, Ann Arbor, 1976; and the articles in A. Montagu (ed.), *Sociobiology Examined*, Oxford University Press, Oxford, 1980; and A. L. Caplan, *The Sociobiology Debate*, Harper & Row, New York, 1978.

[23] Many of these methodological objections can be found in Lewontin (*op. cit.*, 'Sociobiology: A Caricature of Darwinism'), Gould (*op. cit.*, 'Sociobiology: The Art of Storytelling') and Burian (*op. cit.*, 'A Methodological Critique of Sociobiology'). Some examples are: (1) sociobiological theory provides a totally inadequate account of the phenomena to be explained; (2) the genetic and phenotypic variation required by the evolutionary explanations of sociobiologists is simply postulated as needed and any current lack of variation explained away by reference to the variation reducing character of the evolutionary process; (3) a genetic basis for phenomena being explained is assumed and any evidence that might suggest non-genetic transmission (as for example, the incredible short time span – in evolutionary/genetic terms – during which cultural change takes place) is rendered impotent by the postulation of *ad hoc* mechanisms (for example, the multiplier effect), which makes sociobiology immune to evidence and little more than fascinating storytelling; (4) the genome is incredibly more complex than sociobiological theory makes it out to be and the mapping of a genotype onto a phenotype or vice versa is so complex that sociobiological explanations can only

be seen as trivializations which depend for credibility on simplistic and, where necessary to avoid negative evidence, *ad hoc* laws and principles.

[24] See, *op. cit.*, 'A Methodological Critique of Sociobiology', *op. cit.*, 'Sociobiology: The Art of Storytelling', and *op. cit.*, 'Sociobiology: A Caricature of Darwinism'.

[25] Consider, for example, Gould's claim, "Most of *Sociobiology* wins from me the same high praise almost universally accorded to it. For a lucid account of evolutionary principles and an indefatigably thorough discussion of social behavior among all groups of animals, *Sociobiology* will be the primary document for years to come. But Wilson's last chapter, 'From Sociobiology to Sociology,' leaves me very unhappy indeed. After twenty-six chapters of careful documentation for the non-human animals, Wilson concludes with an extended speculation on the genetic basis of supposedly universal patterns of human behavior" (*op. cit.*, 'Biological Potential vs. Biological Determinism', p. 344).

[26] For example, it is not clear that there is a phenomenon to be explained. Unless all (or, at least most) homosexuals have exclusively homosexual encounters, there is no phenomenon of persistence of an apparently deleterious behavior to be explained sociobiologically. If even 25% of dispositional homosexuals engaged in heterosexual activity frequently enough to leave offspring there would be no mystery about the persistence of the behavior, even granting its genetic basis. Hence, given that there is no clear evidence on this matter, sociobiologists cannot be sure that an evolutionary explanation is necessary or appropriate. In addition, there are good reasons for believing that there are considerable social influences on homosexual behavior since, as Lewontin has pointed out, there is known to be an "immense variation in the frequency of homosexual and heterosexual behavior in history and between social classes" (*op. cit.*, 'Sociobiology: A Caricature of Darwinism', p. 28).

[27] See D. S. Saunders, *An Introduction to Biological Rhythms*, Blackie, Glasgow, 1977; and M. C. Moore-Ede *et al., The Clocks that Time Us: Physiology of the Circadian Timing System*, Harvard, Cambridge, 1982.

[28] *Op. cit., Sobiobiology: The New Synthesis*, p. 251.

JAMES H. FETZER

SCIENCE AND SOCIOBIOLOGY*

ABSTRACT. This paper focuses upon the logical structure, the theoretical vocabulary, and the empirical testability of sociobiology. It begins with an informal description of its fundamental theses, together with consideration of its controversial character. These theses are given a semi-formal formulation, and sociobiology, thus summarized, is examined from the perspective of various accounts of scientific theories. Adopting (an intensional version of) a conception first advanced by Campbell sixty years ago, a formalization is provided to illustrate the logical structure of sociobiological laws. The extent to which sociobiology properly qualifies as "scientific" is finally assessed.

Perhaps no question has so long endured within the theory of science as that of establishing an appropriate *criterion of demarcation* on the basis of which, in principle, 'science' ought to be distinguished from 'non-science'. There have been numerous candidates for this position, including principles of verifiability, of falsifiability, and of testability, which have exerted their influence (for better or for worse) upon the great debates that have arisen within the natural and the social sciences during the 20th century. These disputes – many of which continue to this day – have included those of the scientific status of special and of general relativity (within physics), of Darwinian and of Larmarckian evolution (within biology), and of Watsonian behaviorism and of Freudian psychoanalysis (within psychology). When posed against this historical background, therefore, current interest in the scientific standing of sociobiology apparently should generate slight surprise.

The intensity of this debate, however, tends to indicate that sociobiology is not "just another science". There seem to be (at least) three different possible explanations for this state of affairs, relative to members of the scientific community, namely: (a) perception of sociobiology as unscientific (because of some inherent incapacity to satisfy the standards of scientific procedure on logical, methodological or epistemological grounds); (b) rejection of its interdisciplinary character (as though neither natural scientists nor social scientists were willing to allow the very possibility of a "bridge discipline" of this kind); or (c) resistance to its revolutionary significance (where neither "normal" social scientists nor "normal" natural scientists are willing to come to grips with its "paradigm shattering" implications). Indeed, the purpose of this inquiry is to suggest that there are no suitable

James H. Fetzer (ed.), Sociobiology and Epistemology, 217–246.

grounds for contending that sociobiology is not a scientific discipline, in which case perhaps "the sociobiology debate" should be viewed as rooted more in the sociology of science than in the logic of its methods.

1. WHAT IS SOCIOBIOLOGY?

The area of inquiry (or "discipline") known as *sociobiology* is largely the creation of the evolutionary biologist, Edward O. Wilson, who has authored several books, such as *The Insect Societies* (1971), *Sociobiology: The New Synthesis* (1975), *On Human Nature* (1978), and, more recently, in collaboration with physicist Charles R. Lumsden, *Genes, Mind, and Culture* (1981) and *Promethean Fire* (1983). The primary thesis of these works is that many facets of social organization, including aspects of culture itself, as well as patterns of individual development are species specific, i.e., depend upon and vary with particular genetic types, where these cultural characteristics, in turn, affect the capacity of each species to survive and reproduce. From Wilson's perspective, in other words, were we "zoologists from another planet completing a catalog of social species on Earth, . . . anthropology and sociology together [would] constitute the sociobiology of a single primate species", namely: *Homo sapiens* (1980, p. 271). Wilson thus wants to assimilate the study of culture within the framework of evolutionary biology, thereby embracing the program of a natural science of social behavior.

It should not be surprising that this approach has stimulated enormous controversy with scientists and non-scientists alike, not least of all for taking seriously the idea that human culture might be "on a par" with other species' cultures, rather than, say, "the inevitable outcome of either advanced social life or high intelligence", much less of divine intervention (1978, p. 22). Wilson, after all, presumes that the cultures of different species are properly compared, offering the following attributes as illustrating features of every human culture known to history or to ethnography:

> age-grading, athletic sports, bodily adornment, calendar, cleanliness training, community organization, cooking, cooperative labor, cosmology, courtship, dancing, decorative art, divination, division of labor, dream interpretation, education, eschatology, ethics, etiquette, faith healing, family feasting, fire making, folklore, food taboos, funeral rites, games, gestures, gift giving, greetings, hair styles, housing, hygiene, incest taboos, inheritance rules, joking, kin groups, language, law, magic, marriage, mealtimes, medicine, obstetrics, . . . (1978, p. 22)

and dozens more, each of which might be viewed as an aspect of human nature. Indeed, Wilson boldly invites us to contemplate the possibility of even more complex societies whose non-human members are possessed of superior intelligence, yet which differ from human societies in most of these respects. "If ants were endowed . . . with rationalizing brains equal to our own", he says, "they could be our peers". Ant societies could display features like these:

> age-grading, antennal rites, body licking, calendar, cannibalism, caste determination, colony-foundation rules, colony organization, cleanliness training, communal nurseries, cooperative labor, cosmology, courtship, division of labor, drone control, education, eschatology, ethics, etiquette, euthanasia, fire making, food taboos, gift giving, government, greetings, grooming rituals, hospitality, housing, hygiene, incest taboos, language, larval care, law, medicine, metamorphosis rites, mutual regurgitation, nursing casts, queen obeisance, surgery, . . . (1978, p. 22)

and dozens more, each of which might then be taken as an aspect of ant nature. From this point of view, therefore, culture itself is by no means a peculiarly human prerogative, but may also belong to birds, to bees, and to chimpanzees.

In extending the concept of culture to species other than *Homo sapiens*, of course, Wilson tacitly extends its meaning, which typically signifies "the language, habits, skills, arts, instruments, institutions, etc. of a certain people in a certain period", where these attainments are more often supposed to be the products of learning than the offspring of genetic endowment. Yet the scope of scientific formulations must be permitted to exceed the nuances of ordinary language, which inevitably occurs when properties that were previously thought to be exclusively within the province of man are discovered to be among the possessions of other species as well. Moreover, even in its ordinary sense, "culture" as a product of learning tends to presuppose differing capacities for attainment as a function of genetic endowment, in the form, let us say, of "predispositions", i.e., of genetically-based limitations upon the extent to which individuals or collections of individuals have the ability to acquire and to exercise language, habits, skills, arts, instruments, institutions, etc., as functions of species-specific genetically based capabilities.

In *Genes, Mind, and Culture*, Lumsden and Wilson attempt to emphasize the role of mental activity (or of "mind") in the propagation of culture, wherein

> the genes prescribe a set of biological processes, which we call *epigenetic rules*, that direct the assembly of the mind. This assembly is context dependent, with the epigenetic rules feeding on information derived from culture and physical environment. Such information is forged into cognitive schemata that are the raw materials of thought and and decision. Emitted behavior is just one product of the dynamics of the mind, and

> *culture is the translation of the epigenetic rules into mass patterns of mental activity and behavior*. ... We define culture in the broad sense to include the sum total of mental constructs and behaviors, including the construction and employment of artifacts, transmitted from one generation to the next by social learning. (1981, pp. 2–3; italics added)

Thus, insofar as mental activity contributes to the production and utilization of cultural artifacts (including language, habits, skills, arts, instruments, institutions, etc.), which are transmitted to successive generations in part through a process of social learning, then to the extent to which these cultural capabilities tend to promote or to inhibit the evolution of a species, its reproduction and survival is a manifestation of *gene-culture coevolution*. But it should be obvious that gene-culture coevolution can take place only if a species possesses the capability for the transmission of cultural artifacts from one generation to the next by means of social learning, which must thus presuppose "mind" as the cognitive capacity for thought and decision. Hence, to the extent to which a species does not possess these cognitive capacities, it cannot participate in gene-culture coevolution; conversely, to the extent to which a species can participate in gene-culture coevolution, its members must possess the required cognitive capacities, i.e., they must have "minds".

Upon initial consideration, the introduction of "epigenetic rules" would appear to needlessly complicate an otherwise elegent approach. The conception advanced by Wilson's *On Human Nature*, for example, holds that different species S^1, S^2, ... are characterizable by different patterns of social behavior, B^1, B^2, ... : things that are ants exhibit age-grading, antennal rites and so on, while things that are human exhibit age-grading, athletic sports, and so forth. Indeed, a parallel conception within natural science holds that different elements E^1, E^2, ... are characterizable by different combinations of physical attributes, A^1, A^2, ... : things that are gold exhibit a melting point of 1063°C, a boiling point of 2600°C, and so on, while things that are titanium exhibit a melting point of 1800°C, the specific gravity of 4.5, and so forth. Thus, insofar as the physical and chemical properties of the elements recur in a regular fashion when the elements are arranged in increasing order of their atomic number (as reflected by the *Periodic Table of the Elements*), possibly the personal and social properties of the species might be found to recur in a regular fashion when the species are arranged in some order appropriate to their genetic origins (as reflected by a new *Periodic Table of the Species*).

An underlying difficulty appears to accompany this conception, however, because 'patterns of social behavior' are not properties of ants and humans in precisely the way that 'combinations of physical attributes' are properties of gold and titanium: things that are gold, for example, possess their melting points and boiling points as properties of discrete individual things, whereas things that are human, by contrast, possess their age-grading and athletic sports, not as properties of discrete individual things, but instead as relations between various individual things. Thus, while physical attributes are properties of individual things *distributively* (since each thing that is gold has a melting point of 1063°C without regard to its relations to other things that are gold), social behaviors are properties of individual things *collectively* (since age-grading, for example, is not a property that one human has as a discrete individual thing, but only exists as a relation between that human and other humans). And, if this is the case, then, in order to provide an essential mechanism for the generation of social behavior on the basis of properties of discrete individual things, "epigenetic rules" would need to be introduced to represent the potential for social behavior as a dispositional property of discrete individual things that manifests itself under the relevant social conditions, roughly as *Genes, Mind, and Culture* tends to suggest.

Epigenetic rules, in other words, exist in the form of predispositions to acquire one or another specific behavioral tendency BT^1, BT^2, and so on, within a fixed range of specific social patterns B^i, where which among these potential behavioral tendencies happen to be acquired as actual dispositions depends upon and varies with environmental factors, such as social learning. The variety of specific behavioral tendencies that an individual acquires as actual dispositions collectively reflects its "selectivity", while the strength of that individual's tendency to manifest one or another of those actual dispositions distributively reflects its "penetrance"; thus, an individual's actual behavioral tendencies during any particular moment in its development may be described by corresponding *usage bias curves* (Lumsden and Wilson, 1981, pp. 55–65). Thus, while *Homo sapiens* has the predisposition toward athletic sports, this tendency may be manifest by acquiring any of a broad variety of specific dispositions to play soccer, tennis, and so forth, with distinctive strengths of those tendencies relative to different environmental conditions, where which, if any, among these happens to be displayed by a particular member of the species is partly a function of social learning. Yet ants, by way of comparison, are completely incapable of acquiring the disposition to play soccer, or tennis, and so on, precisely because they lack the predisposition toward athletic sports that numbers among the characteristics of human genes.

From this point of view, therefore, sociobiology addresses the necessity to assess the relative influence of genes and of culture (or of "nature" and of "nurture") on the social behavior displayed by different species, where if social learning factors are comparatively unimportant in the emergence of various behavioral tendencies, then those specific traits are properly classified as more-or-less "biologically determined", but otherwise not. Thus, the major difference between alternative accounts of sociobiology is likely to arise at just this juncture; for, as Wilson has observed, sociobiology, in general, is compatible with (at least) three different kinds of relations between genetic predispositions and their behavioral manifestations (cf. Wilson, 1978a, p. xi):

(a) every member of the species possesses the same genetic predisposition, and this predisposition is compatible with a broad range of possible behavior;
(b) every member of the species possesses the same genetic predisposition, but this predisposition is compatible with a narrow range of possible behavior;
(c) some members of the species of possess different genetic predispositions, where these predispositions are compatible with different ranges of possible behavior, some of which are broad and some of which are narrow, respectively.

Although Wilson was not initially committed to one or another of these theoretical alternatives, the genetical, the psychological and the anthropological evidence led him to conclude "that human social behavior is to some extent genetically constrained over the entire species and furthermore [is] subject to genetic variation within the species", i.e., to accept theses (c):

> The most emotion-laden and least rationally controlled human behaviors are generally consistent in their details with sociobiological theory. These categories include incest inhibition, bond formation, parent-off-spring conflict, sex-biased infanticide, primitive warfare, territoriality, and sexual practice. They are more simply explained by the hypothesis of genetically based predisposition than by the hypothesis of purely cultural determination. (Wilson, 1978a, p. xii)

Sociobiology thus provides a framework for explaining the social behavior of different species as a (complex) function of genetic and of cultural factors, where some species will tend to exhibit uniform genetic predispositions, most of which permit only narrow ranges of possible behavior ("the lesser species"), but other species will tend to exhibit variable genetic predispositions, many of which permit far broader ranges of possible behavior ("the higher species").

Although sociobiology has not met with unqualified enthusiasm from every quarter, it has sometimes been subjected to excessive abuse. In his scathing review of *Genes, Mind, and Culture*, for example, Edmund Leach remarks, "This book comes so close to being a parody of the genre to which it belongs that I have had difficulty in believing that it is not intended as an academic hoax" (Leach, 1981, p. 267). Acknowledging that Wilson has made important contributions to understanding the social behavior or insects, Leach rakes Lumsden and Wilson over the coals for committing a wide variety of (alleged) theoretical mistakes and factual blunders, not least of which is their assumption, "that human culture can be broken down into clusters of traits (which the authors call 'culturgens') which can then be counted and subjected to statistical analysis". In Leach's view, once Lumsden and Wilson's needless jargon has been decoded, their basic claim comes to no more than that man's genetic endowments make it more likely that some aspects of human culture will be more developed than will other aspects, which Leach regards as a matter of scant importance:

> The human species is not genetically uniform but the variations seem to be minor. No correlation has yet been established between known genetic variations and morphological variations [of form and structure]. Equally, there is no clear correlation between morphological variations (the so-called "racial types"), cultural variations and/or modes of subsistence. But equally there are no cultural universals except at the level of such generalizations as "all human beings have speech" [and] "all human societies distinguish between domestic space and non-domestic space".

Thus, while Leach agrees with Lumsden and Wilson that the findings of biology and of social science need to be compatible for either to rate as "science", the discipline known as *sociobiology*, in his estimation, does not measure up.

However plausible Leach's critique may appear upon initial consideration, it would be quite wrong to imagine that Leach has had the last word. To begin with, if the findings of biology were inconsistent with the findings of social science, that in itself surely would not suffice to establish that either were *unscientific*, even though they could not then both be altogether true: classical mechanics, after all, is incompatible with special relativity, yet nevertheless qualifies as "scientific". Indeed, Karl R. Popper, especially, holds that falsifiability is the appropriate criterion of demarcation for separating science from non-science; thus, by Popper's standard, theories that are known to be false are thereby proven to be *scientific*. Even if Leach were right to deny the existence of correlations "between known genetic variations and morphological variations", he could at most have shown that sociobiology is not true, but not that it is "unscientific". Moreover, by denying relationships of this kind, Leach implies that morphological variations between Watusi and Pygmy, between Caucasians and Orientals, etc., are not rooted in their genes. The evidence suggests not that sociobiology is false but that Leach is wrong.

Furthermore, if the human species were genetically uniform to a high degree, then (presumably) it would be less difficult to discover the very generalizations to which sociobiology aspires than would otherwise be the case, because observed variations in social practices (almost certainly) would not reflect unobserved variations in genetic structures. Under these conditions, observed variations in social practices would support an inference regarding the variety in range of manifestations permitted by a single social genotype, rather than regarding different social genotypes, each of which permits only a single variety of manifestations. Whether or not Leach would be willing to acknowledge these differences, his position here seems to be either question-begging or simply false. Of course, if there were no "cultural universals", or if cultural traits could not be identified and subjected to some sort of mathematical depiction, sociobiology would be hopeless. Leach, however, not only fails to show that social practices (such as age-grading, athletic sports, and so on) cannot be described and their instances counted (as numbers of, say, games of soccer, sets of tennis, and so forth), but he also offers examples of "culturgens" of his own; for, if all human beings have speech and if all human societies distinguish domestic from non-domestic space, then there must be at least some "cultural universals", where the issue becomes a disagreement over their variety and distribution. Leach thus appears to have been carried away in assaulting this area of inquiry, for the objections that he has put forth either miss their mark or prove too weak to establish his own "conclusions".

By defending Lumsden and Wilson from these unwarranted criticisms, however, I do not intend to suggest that all their formulations are necessarily free from fault. Still, their problems may not lie where they initially appear to lie. Wilson's fanciful society of superior ants, for example, might invite ridicule, insofar as ant-cosmology, ant-law, and ant-medicine may be implausible in the extreme, at least to the untutored. Indeed, the reasoning he employs in constructing this scenario is itself subject to complaint, since one might as well conclude, with similar justification, that if ants were endowed with rationalizing brains equal to our own, then their bodies could not support their heads, their average weight would be 175 lbs., and their colonies would occupy the space of mountains. But caution is required here, since the tricky part about counter-factual conditionals is figuring out what to keep constant as well as what to change: the invention of silicon chips by the computer industry conclusively shows that things that are small can be things that are smart! Over 8,000 species of ants have so far been described and classified, with the estimate that at least another 4,000 remain to be discovered, less than 1% of which have been studied in detail (Wilson, 1980, p. 189). If evidence were to warrant the rationality of ants as "an inference to the best explanation", moreover, then that very hypothesis – together with its consequences for anti-cosmology, ant-cosmology, anti-law and ant-medicine – would deserve tentative acceptance (Fetzer, 1981, pp. 265–273). The existence of *Antus superbus*, therefore, is not purely an abstract possibility.

From a philosophical point of view, of course, the serious problems with Lumsden and Wilson (1981) are theoretical questions rather than factual ones; for even if Leach were entirely right in objecting to their claim that, "the advance of men over chimpanzees and other higher primates was attained during an exceptionally rapid burst of evolution", as insufficiently supported by available evidence (cf. Lewin, 1982), this would not justify the inference that their undertaking is completely without value. Their general strategy, after all, is to discover the relationship between genetic origins (as instances of different genotypes G^1, G^2, . . .) and bodily structures (as instances of different phenotypes P^1, P^2, . . .) under the influence of environmental factors (as instances of different physical and social factors EF^1, EF^2, . . .), where these phenotypes, in turn, are predisposed toward the manifestation of different patterns of individual and social behavior (as instances of different epigenetic rules ER^1, ER^2, . . .), while the individual and social behavior that arises from the influence of environmental factors (as instances of different physical and social factors EF^1, EF^2, . . .), in turn, affects the evolution of the species (as instances of different rates of reproduction and survival r^1, r^2, . . .). Problems that might arise with the fundamental concepts that are involved here, such as genotype, phenotype, and epigentic rule, therefore, are enormously significant. When Leach attacks their conception of epigenetic rules as unclear and imprecise, in other words, he raises a far more threatening issue than would arise if there should happen to be no discernable correlation between, say, an individual's cranial size and his intellectual capacity.

Objections of this kind are fundamental, because whether or not sociobiology properly qualifies as a scientific discipline depends far more upon its theoretical formulations and their empirical testability than upon the truth or the falsity of its claims. Scientific methods of investigation are characterizable by adherence to (what Popper has called) "the principle of empiricism", namely: that the acceptance or the rejection of hypotheses or theories is determined by observation and by experimentation (Popper, 1968, pp. 54–55); yet the discovery that some specific formulation of a sociobiological generalization might turn out to be false surely should not be viewed as appropriate grounds for abandoning the sociobiological program, since some other formulation of that same generalization might turn out to be true. An important distinction has to be drawn between the *scientific acceptability* of an hypothesis or a theory and the *scientific rationality* of pursuing the "research program" that they happen to represent, more or less as Lakatos has emphasized; for an hypothesis or a theory should be tentatively rejected when its truth is incompatible with the available relevant evidence, relative to acceptable rules of inference, but the rejection of one hypothesis or theory does not mean that another hypothesis or theory of the same or of a similar kind cannot fare better.

So long as sociobiologists remain willing to reject specific formulations of sociobiological hypotheses and theories in accordance with the principle of empiricism, the conclusion that sociobiology is unscientific should not follow merely because some of their formulations are false, any more than the conclusion that sociobiology is scientific would follow merely because some of their formulations are true. A description of the contents of my pocket might well be true, but it would not qualify as scientific. If sociobiology is not scientific, therefore, then it must be because the formulations representing, expressing, or conveying its (basic) hypotheses and theories are not acceptable, either because they are too vague, too ambiguous, empirically untestable, or otherwise fail to meet scientific standards. The question, in other words, is not that of the scientific acceptability of sociobiology as an hypothesis or as a theory that is true, but rather of the acceptability of sociobiology as a scientific hypotheses or as a scientific theory, no matter whether true of false. For if the language in which sociobiological hypotheses and theories are characteristically couched could not satisfy appropriate standards of scientific acceptability, then the pursuit of sociobiology would not qualify as a scientific program, since its hypotheses and theories would lack empirical content – if not altogether then at least to a large degree (cf. Hempel, 1965).

Indeed, these potential shortcomings are by no means wholly unrelated to one another, since vague hypotheses (such as, "Something good will happen in your life") tend to be compatible with almost *any* possible outcome (such as, say, finding a quarter on the sidewalk, receiving a letter from a friend, or being elevated to high position), which renders them virtually invulnerable to falsification; while ambiguous hypotheses (such as, "All men are created equal") might be *false* under one interpretation (as a descriptive claim concerning the distribution of human abilities) but *true* under another (say, as a normative ideal regarding equal justice under law). The evidence suitable for testing an ambiguous hypothesis under one interpretation, therefore, may be drastically different than the evidence suitable for testing that same hypothesis under another interpretation, where appropriate investigations can be pursued only when those ambiguities have been disclosed and displaced by more specific language. Thus, if the vocabulary of sociobiology, including "species", "genotype", "phenotype", and "culturgen", is vague and ambiguous or otherwise unsatisfactory, then its scientific status will stand in doubt. To the extent to which science aims at explanatory hypotheses and theories, moreover, sociobiology would remain unscientific if it could not provide appropriate explanations for the phenomena to which it applies, even were its vocabulary wholly composed of precisely defined and unambiguous constructs. Let us therefore consider whether or not sociobiology has the right sort of general structure and technical vocabulary to properly qualify as "scientific".

2. WHEN ARE THEORIES SCIENTIFIC?

In order to approach this issue in a systematic fashion, we will begin by formulating the fundamental principles of sociobiology as a sequence of theses of two different kinds, namely: *definitions*, which reflect the meanings to be assigned to technical terms in the vocabulary of sociobiology; and *conjectures*, which reflect the principal relations that are supposed to obtain between these properties, thus understood, if sociobiology happens to be true. "Definitions", therefore, are assumed to be true on syntactical or on semantical grounds alone, while "conjectures" are logically contingent empirical assertions that are true if what they assert to be the case is the case; otherwise, they are false. We shall then consider whether one or another conception of the logical character of scientific theories sheds light on the scientific standing of sociobiology.

Let us begin with deterministic versions of all of these definitions and conjectures – including the laws of Mendelian genetics – and subsequently consider the formulations that would be required of indeterministic counterparts. This desideratum should be borne in mind regarding what now follows.

Among the most basic assumptions underlying sociobiology is the following:

CONJECTURE 1. Phenotypes develop from genotypes under the influence of environmental factors.

Comment. This conjecture is a thesis of molecular genetics and of evolutionary biology, which is presupposed by sociobiology. Strictly speaking, *instances* of phenotypes develop from *instances* of genotypes under the influence of *instances* of (physical and social) environmental factors. The thesis has two corollaries:

COROLLARY 1A. Different phenotypes P^1, P^2, . . . must develop from the same genotype G under the influence of different environmental factors EF^1, EF^2, . . . ;

COROLLARY 1B. Different phenotypes P^1, P^2, . . . must develop from different genotypes G^1, G^2, . . . under the influence of the same environmental factors EF.

Comment. So long as "genotypes" are defined in terms of genetic material and not in terms of their predispositions to develop into "phenotypes" (defined in terms of neurological and physiological structure), these theses are empirical assertions that are logically contingent. For claims of this kind to be true, the specification of genotype and of environmental factors must include the presence or the absence of every property whose presence or absence makes a difference to the development of a corresponding phenotype, while causally irrelevant properties are excluded.

CONJECTURE 2. Epigenetic rules are properties of phenotypes.

Comment. This conjecture is the fundamental thesis of general sociobiology. Strictly speaking, *instances* of (sets of) epigenetic rules are properties of *instances* of phenotypes, where these properties vary according to phenotype:

COROLLARY 2A. Every instance of the same phenotype P must possess the same (set of) epigenetic rules ER^1, ER^2, . . . ;

COROLLARY 2B. Different (sets of) epigenetic rules ER^1, ER^2, . . . cannot belong to instances of the same phenotype P.

Comment: Corollary 2B is merely the contraposed form of Corollary 2A, which is added for emphasis. Note, in particular, that every instance of the same phenotype must have the same epigenetic rules, if sociobiology is true, even though the truth of this assertion does not obtain on logical grounds alone.

DEFINITION 1. Epigenetic rules are predispositions to acquire and to manifest one or another specific behavioral tendency within a fixed range of possible patterns of (individual and social) behavior, under the influence of (physical and social) environmental factors.

Comment. The possession of a specific epigenetic rule determines the breadth and variety of potential patterns of behavior that an instance of a phenotype can acquire and can manifest, but that range itself might be broad or narrow.

DEFINITION 2. The narrower the range of possible patterns of behavior B^i permitted by a specific epigenetic rule ER^i, the smaller the number of specific behavioral tendencies BT^1, BT^2, ... that can be acquired or manifest.

DEFINITION 3. The broader the range of possible patterns of behavior B^i permitted by a specific epigenetic rule ER^i, the larger the number of specific behavioral tendencies BT^1, BT^2, ... that can be acquired or manifest.

Comment. Strictly speaking, the narrower the range of possible patterns of behavior B^i permitted by an *instance* of a phenotype P possessing a specific epigenetic rule ER^i, the smaller the number of specific behavioral tendencies BT^1, BT^2, ... that can be acquired or manifest (as actual dispositions) by that *instance* of that phenotype under the influence of (physical and social) environmental factors. Definition 3 is a trivial complement of Definition 2.

DEFINITION 4. Cultures are composed of culturgens.

Comment. This concept applies to sociobiology and to cultural anthropology. Strictly speaking, *instances* of cultures are composed of *instances* of culturgens as tokens of various types, which can be counted, classified, and so on.

DEFINITION 5. Culturgens consist of artifacts (including tools, clothing), of behavior (such as speech, skills), and of mentifacts (including thoughts and ideas).

DEFINITION 6. Epigenetic rules predispose phenotypes toward the acquisition and manifestation of specific behavioral tendencies BT^1, BT^2, . . . involving culturgens.

Comment. Strictly speaking, *instances* of phenotypes possessing various kinds of epigenetic rules are predisposed toward the acquisition and the manifestation of specific behavioral tendencies BT^1, BT^2, . . . involving the utilization of specific *instances* of culturgens, say, tools of a certain kind, particular forms of speech, etc., which may occur at different times and places.

CONJECTURE 3. Culturgens can be transmitted by social learning.

Comment. This thesis is likewise fundamental to general sociobiology. Note, in particular, that this is a logically contingent empirical assertion, which would be false if culturgens were never transmitted by way of social learning.

COROLLARY 3A. Every instance of the same phenotype *P* must be predisposed to acquire and to manifest the same culturgens under the same conditions of social learning.

COROLLARY 3B. Different predispositions to acquire and to manifest the same culturgens under the same conditions of social learning must be instances of different phenotyes.

Comment. Corollary 3B is merely the contraposed form of Corollary 3A, which is added for emphasis. Note, in particular, that while every instance of the same phenotype must have the same social learning predispositions, that would remain the case even if that social learning predisposition were non-existent, so long as the same predisposition is possessed by every instance of the same phenotype with the same strength, no matter what its degree of strength may be.

CONJECTURE 4. Phenotypes possessing the same epigenetic rules may nevertheless acquire and manifest different specific behavioral tendencies.

Comment. This conjecture, which is enormously important to sociobiology, must be carefully distinguished from Conjecture 2. That every instance of the same phenotype must possess the same (set of) epigenetic rules does not entail that every instance of the same phenotype must acquire and manifest the same specific behavioral tendencies, unless they are under the influence of the same sets of (physical and social) environmental factors at every moment of their lives.

COROLLARY 4A. Every instance of the same phenotype P must acquire and manifest the same specific behavioral tendencies $BT^1, BT^2, \ldots$ if it is subjected to the same (set of) environmental factors $EF^1, EF^2, \ldots$;

COROLLARY 4B. If instances of the same phenotype P acquire and manifest different specific behavioral tendencies $BT^1, BT^2, \ldots$, then they must have been subjected to different (sets of) environmental factors $EF^1, EF^2, \ldots$.

Comment. Since instances of the same phenotype P must possess the same epigenetic rules, according to Conjecture 2, yet may acquire and manifest different specific behavioral tendencies $BT^1, BT^2, \ldots$, nevertheless, this can occur *only if* those instances were subject to different (sets of) environmental factors $EF^1, EF^2, \ldots$ at some moment of their lives, if these theses are true.

DEFINITION 7. Phenotypes have minds to the extent to which their epigenetic rules are under the influence of social learning.

Comment. This conception, although controversial, appears to be a thesis of *Genes, Mind, and Culture*, at least insofar as Lumsden and Wilson embrace the conception of gene-culture co-evolution described in Section 1, according to which (the members of) a species can participate in gene-culture co-evolution only if they possess the capacity for the transmission of specific behavioral tendencies, especially concerning the utilization of culturgens, by means of social learning. This definition may be viewed as open to re-interpretation.

DEFINITION 8. (Sexual) species are populations of phenotypes that are capable of interbreeding freely under natural – voluntary and unforced – conditions.

DEFINITION 9. Societies are (collections of) the members of a species that are organized in a fashion that promotes communication and cooperation beyond what is required to pursue sexual activities.

Comment. These definitions belong to biology, to sociobiology, and to anthropology. Their significance arises from additional theses in which they occur.

CONJECTURE 5. Societies are manifestations of the specific behavioral tendencies that their members have acquired as instances of species (phenotypes) under the influence of environmental factors.

COROLLARY 5A. Societies of the members of species whose epigenetic rules permit only a narrow range of possible patterns of behavior can acquire and manifest only a small number of behavioral responses to different environmental conditions.

COROLLARY 5B. Societies of the members of species whose epigenetic rules permit somewhat broader ranges of possible patterns of behavior can acquire and manifest a larger number of behavioral responses to different environmental conditions.

Comment. The principal differences between distinct species, therefore, consists in the breadth and variety of those fixed ranges of possible behavioral tendencies that their members can acquire and manifest as actual dispositions under the influence of (physical and social) environmental factors: the more narrow the range and the less subject to social learning, the greater the justification for regarding specific behavioral tendencies as "instinctual" or as "biologically determined", even though those tendencies invariably arise under the influence of (physical and social) environmental factors themselves. And, while these corollaries have consequences for the evolution of species (since broader ranges of possible response to different environmental conditions confers a selective advantage upon a species within a changing environment), what has been said should be enough to enable us to pursue the matter before us.

The issue of whether or not sociobiology may be cast in the form of a scientific theory, of course, crucially depends upon what form "scientific theories" are supposed to possess. In an attempt to formulate his response to the question, "Is biology different from physics?", for example, Michael Ruse tentatively adopts the "standard conception" of scientific theories as deductive systems associated with R. B. Braithwaite (1953) and with Carl G. Hempel (1965). According to this analysis, a scientific theory is composed of an *abstract calculus* (with a vocabulary of logical and of non-logical elements, a set of formation rules determining its well-formed formulae, and a set of transformation rules determining what follows from what) coupled with an *empirical interpretation* of at least some of its primitive or defined non-logical elements, which infuses it with empirical content. Ruse assumes that if some portion of biology, such as population genetics, can be cast in the same form as some portion of physics, such as statistical mechanics, that portion of biology will thereby have been shown to be "scientific" (Ruse, 1977). Strictly speaking, this procedure, in principle, can establish no more than that some portion of biology has been shown to be "as scientific as" some portion of physics on the basis of (what is supposed to be) a suitable standard; such demonstrations, therefore, have the logical properties of relative consistency proofs as they occur in geometry, for example, where an hyperbolic geometry might be proven to have a Euclidean model, in which case the first is consistent if the second is consistent, as Ruse, no doubt, must be aware.

The problem with a "relative scientificity proof", as it might be referred to, is that the standard conception of scientific theories upon which it is based may or may not qualify as a suitable standard. Consider, as an illustration, the very simple abstract calculus consisting of a vocabulary of four predicate variables W, X, Y, and Z, universal and temporal quantifiers, (x) and (t), and the material conditional connective, $\supset$, assuming ordinary formation and transformation rules, such that the following three formulae,

(I) 1. $(x)(t)(Wxt \supset Xxt)$
2. $(x)(t)[Xxt \supset (Yxt \supset Zxt^*)]$
3. $(x)(t)[Wxt \supset (Yxt \supset Zxt^*)]$

(where t^* is equal to or later than t), constitute a deductive system, where 1 and 2 may be viewed as assumptions (or as "axioms") and 3 as a consequence (or as a "theorem"). Then, in order for this set of formulae to qualify as a scientific theory, it is only necessary to assign an empirical interpretation for (at least some of) the four predicate variables, W, X, Y, and Z, so they will stand for specific classes and/or properties of the physical world.

In relation to the informal summary of sociobiological theses presented above, it is easy to show that sociobiology can satisfy this condition. Let us take Conjecture 1, relating phenotypes P^1, P^2, ... to epigenetic rules ER^1, ER^2, ... and Definition 1, relating epigenetic rules, in turn, to the acquisition and manifestation of specific behavioral tendencies BT^1, BT^2, ... under the influence of (physical and social) environmental factors EF^1,

$EF^2, \ldots$, for example, as "assumptions", where W, X, Z, and Y are assigned corresponding interpretations, respectively, in relation to formulae (I):

$$\begin{aligned}
\text{(II)}\quad & 1.\ (x)(t)(P^1xt \supset ER^1xt)\\
& 2.\ (x)(t)[ER^1xt \supset (EF^1xt \supset BT^1xt^*)]\\
& 3.\ (x)(t)[P^1xt \supset (EF^1xt \supset BT^1xt^*)]
\end{aligned}$$

which, informally expressed, assert, "If x is of phenotype P^1 at t, then x will possess epigenetic rule ER^1 at t", "If x possesses epigenetic rule ER^1 at t, then if x is under the influence of environmental factors EF^1 at t, x will acquire or manifest specific behavioral tendency BT^1 at t^*", and "If x is of phenotype P^1 at t, then if x is under the influence of environmental factors EF^1 at t, then x will acquire or manifest specific behavioral tendency BT^1 at t^*", respectively, as "axioms" and as "theorems" of sociobiology.

Unfortunately, while sociobiology easily fulfills such a requirement, so does almost any other interpretation of the variables in question. Consider conjectures concerning Boy Scouts, for example, who are supposed to be trustworthy, loyal, courteous, kind, obedient, cheerful, thrifty, brave, clean and reverent, which we may interpret as descriptive rather than as normative designations. Since individuals who are trustworthy TW return excess change RC when they receive excess change EC (among innumerable other manifestations of this specific trait), let us formulate another interpretation of formulae (I):

$$\begin{aligned}
\text{(III)}\quad & 1.\ (x)(t)(BSxt \supset TWxt)\\
& 2.\ (x)(t)[TWxt \supset (ECxt \supset RCxt^*)]\\
& 3.\ (x)(t)[BSxt \supset (ECxt \supset RCxt^*)]
\end{aligned}$$

which, formally expressed, respectively assert, "If x is a Boy Scout at t, then x is trustworthy at t", "If x is trustworthy at t, then if x receives excess change at t, x will return that excess change at t^*", and "If x is a Boy Scout at t, then if x receives excess change at t, then x will return it at t^*". Yet even if these assertions were true, no one should imagine that (III) affords a scientific explanation of trustworthy behavior among Scouts. While (II) and (III) are empirical interpretations of an abstract calculus, thereby satisfying the standard conception, it is not obvious that this is sufficient for a set of formulae to properly qualify as a scientific theory.

Perhaps, therefore, the difficulty lies with the "standard conception" itself; for it would not be surprising if scientific explanations were not forthcoming from inadequately conceived "scientific theories". Let us then turn to the "semantical conception" of scientific theories as set-theoretical predicates, which is identified with Patrick Suppes (1967) and with Joseph C. Sneed (1971). According to this approach, a scientific theory consists of a *theoretical definition* (of the form, "*x* is a ____-kind of system if and only if *x* satisfies conditions . . .") that may be used to make *empirical claims* (of the form, "Objects *d, e,* . . . are systems of kind ____"). Thus, while scientific theories, strictly speaking, are stipulative definitions (or semantical conventions), they can be employed to describe the physical world. Such a system might be defined as a "classical particle system" if and only if it is a system of physical objects which satisfies the inverse-square law of universal gravitation and Newton's three laws of motion; and the empirical claim could then be advanced that the Sun and planets of our solar system is a classical particle system. Another system might be defined as a "Mendelian hereditary system" if and only if it is a system of sexually reproducing objects which satisfies Mendel's three laws of inheritance; and the empirical claim could then be made that the transmission of eye-color in *Homo sapiens* is a Mendelian hereditary system (cf. Giere, 1979). Yet another might be defined as a "normal driving system" if and only if it is a system of drivers, automobiles, and highways which satisfies the valid-license law, the 55 mph speed-limit law, and the unexpired-auto-tag law; the empirical claim might then be made that Colorado (at least, in part) is a normal driving system.

Clearly, something is wrong, because the sense in which normal driving systems and the sense in which Mendelian hereditary systems involve "laws" are surely not the same. Automobile laws, after all, require passage by a legislature, interpretation by the judiciary, and enforcement by the executive; Mendel's laws, by contrast, require no lawgiver and need no enforcer: natural laws, unlike manmade laws, cannot be violated and cannot be changed. Even if every Boy Scout displays trustworthy behavior, one strongly suspects (a) that their trustworthy behavior has more to do with family child-rearing practices than with the payment of membership fees and (b) that every Scout, alas, also has his price, i.e., that this generalization could be made false by means of some procedure, such as bribes. Similarly, even if every little old lady with gray hair who owns a prize-winning Pomeranian also has a Volks, one may speculate (a) that their automobile ownership has nothing to do with their poodles' achievements and (b) that even some of them might be induced, under specific conditions, to surrender their VW's, i.e., that this generalization likewise could be made false by means of some process, such as, say, the gift of a Mercedes. If these considerations are correct, however, then the limitations of the standard conception may be rooted in its tendency to ignore the logical structure of *lawlike sentences*, i.e., of sentences which would be laws if they were true, a desideratum which the semantical conception treats as though it were something that could be presupposed. Without an adequate analysis of natural laws, in other words, neither of these conceptions of scientific theories could possibly serve as a suitable standard.

The underlying difference between lawlike sentences and (merely) unrestricted generalizations thus appears to be that laws preclude something a (mere) generalization permits, namely: the existence of properties or procedures, natural or contrived, by means of which an attribute, such as being trustworthy, could be taken away from a reference property, such as being a Boy Scout, without taking that reference property away as well, e.g., by bribing a Scout. If such a possibility is permitted, then that generalization, even when true, reflects a *transient property* relation; if such a possibility is precluded, then that generalization, when true, reflects a *permanent property* relation (Fetzer, 1981). As a linguistic manifestation of this ontological difference, therefore, subjunctive conditionals ought to be employed in lieu of material conditionals for these stronger claims, where a subjunctive, $\ni\!\!-$, entails the corresponding material conditional:

(IV) '$p \ni\!\!- q$' entails '$p \supset q$';

that is, if a subjunctive is true, then the corresponding material conditional must be true; but not conversely. Thus, the difference at issue is parallel to the distinction drawn by Aristotle between universal and commensurately universal properties of things (*Posterior Analytics*, I, 73a–74a), where the permanent property relation displaces the Aristotelian essential property relation.

A more adequate conception of scientific theories, therefore, appears to have been advanced by Norman R. Campbell (1920), according to which a scientific theory consists of *lawful hypotheses* conjoined with *dictionary definitions* affording (possibly incomplete) translations of its theoretical vocabulary into empirical language. An analogous conception, which does not presuppose the observational/theoretical language distinction, moreover, has been advanced by the present author, according to which a scientific theory consists of *lawlike sentences* and *meaning postulates* (as definitional sentences that may be either "partial" or "complete") that apply to a common domain (Fetzer, 1981, pp. 156–161). If we will admit an additional intensional operator, $\ni\!\!-_u$, as a "causal conditional" of universal strength u that entails a corresponding subjunctive:

(V) $`p \ni\!\!\rightarrow_u q'$ entails $`p \ni\!\!\rightarrow q'$;

where, if a causal conditional of universal strength is true, then the corresponding subjunctive conditional must be true, but not conversely, the difference between lawlike sentences of simple form and lawlike sentences of causal form can be formally displayed (and the difference between "deterministic" and "indeterministic" scientific theories can later be characterized syntactically). These considerations thus afford a foundation for discriminating between "must" and "will" in the formalization of lawlike relations; for "must", unlike "will", designates non-logical permanent property relations that cannot be transgressed.

The appropriate standard for formalizing a set of formulae as a scientific theory, therefore, requires the invocation of intensional connectives, without which the distinction between lawlike and (merely) "accidental" generalizations cannot be suitably drawn (Fetzer, 1981, pp. 28–35). By acknowledging these differences, it become possible to differentiate between assertions that are intended as lawlike claims and assertions that are not, on the basis of syntactical and semantical criteria. An appropriate formalization of the fragment of sociobiology that is inadequately represented by (II), in other words, can now be adequately represented by appealing to intensional connectives, as follows:

(VI) 1. $(x)(t)(P^1xt \ni\!\!\rightarrow ER^1xt)$
2. $(x)(t)[ER^1xt \ni\!\!\rightarrow (EF^1xt \ni\!\!\rightarrow_u BTxt^*)]$
3. $(x)(t)[P^1xt \ni\!\!\rightarrow (EF^1xt \ni\!\!\rightarrow_u BTxt^*)]$

which, informally expressed, assert respectively, "If x were of phenotype P^1 at t, then x would possess epigenetic rule ER^1 at t", "If x were to possess epigenetic rule ER^1 at t, then x's being under the influence of environmental factors EF^1 at t would (invariably) bring about x's acquiring and manifesting specific behavioral tendency BT^1 at t^*", and "If x were of phenotype P^1 at t, then x's being under the influence of environmental factors EF^1 at t would (invariably) bring about x's acquiring and manifesting specific behavioral tendency BT^1 at t^*". Or, even more informally, "Instances of phenotype P^1 must possess epigenetic rule ER^1", "Individuals that possess epigenetic rule ER^1 must acquire and manifest specific behavioral tendency BT^1 if they are subject to the influence of environmental factors EF^1", and "Instances of phenotype P^1 must acquire and manifest specific behavioral tendency BT^1 if they are subject to the influence of enviornmental factors EF^1", which appear to be the intended interpretations.

The fundamental difference between (II) and (VI), therefore, is that (II), unlike (VI), is not adequate to reflect the lawlike content of sociobiology as informally summarized above. The inherent insufficiency of alternative conceptions, which either fail to acknowledge or else fail to capture the distinction between lawlike and accidental generalizations, in retrospect are apparent; for the acceptability of (III) and the unacceptability of (II) as appropriate formalizations results from the realization that (III), at best, would ordinarily never be entertained as anything more than merely accidental generalizations, while (II) was always intended to be understood as making lawlike claims, as (VI) more appropriately displays. As a result, if (VI) is true, not only will no sociobiologist ever discover an instance of that phenotype that has been under the influence of those environmental factors yet does not manifest the corresponding behavioral tendencies, but no social or natural scientist, no matter what their resources, could succeed in creating an instance of which that were not true. The truth conditions for lawlike claims, therefore, are very strict, indeed; for, in technical terms, their reference class descriptions have to be *maximally specific*, such that for every predicate that designates any property whose presence or absence makes a difference to the presence or the absence of such an attribute, either that predicate F or its negation $-F$ must be included. If a true generalization allows of any exceptions, then it cannot be a law; and if a lawlike generalization allows of any exceptions, then it cannot be true. The crucial issue, therefore, is the character of scientific laws and not the structure of scientific theories; for without a satisfactory analysis of laws, the standard conception and the semantical conception are both clearly inadequate; with such an analysis, however, they are equally obviously superfluous.

3. IS SOCIOBIOLOGY SCIENTIFIC?

Perhaps it should be emphasized that the scientific acceptability of the sociobiological vocabulary does not depend upon whether it can be subjected to operational definition, replacement by observables, or partially defined using reduction sentence "meaning postulates", whatever may be said on their behalf. Hempel, for example, has abandoned the standard conception of scientific theories and has proposed that the characteristic vocabulary and the lawlike sentences that constitute the *internal principles* of a novel theory should be intelligible relative to previously understood concepts and theories, which may serve as *bridge principles* relating innovative formulations to antecedently available experimental procedures, observational techniques or background knowledge (Hempel, 1966). So long as the principle of empiricism retains its force, then this context-dependent, historically-relative pragmatic conception, which seems to do justice to the "theory-laden" character of observational language, would appear to afford a desirable framework for examining issues of this kind. Indeed, there is reason to believe that only pragmatical constructions are adequate, in principle, to capture the conception of a "language-using community". From this point of view, if (portions of) the vocabulary of sociobiology can be rendered intelligible in relation to those of antecedently understood formulations, such as those of molecular genetics and of evolutionary biology, surely they should qualify as "scientifically significant" in some appropriate sense.

As a result, if "genotypes" can be understood by reference to DNA, alleles and genomes, if "phenotypes" can be understood by reference to anatomy and neurology, and if "species" can be understood by reference to Mendelian laws and population genetics, then sociobiology may build upon a foundation laid by molecular genetics, morphology, and evolutionary biology. The difficulty encountered here, however, is already quite familiar; for these elements of the sociobiological vocabulary may thereby be shown to be not "scientific" but "as scientific as" corresponding elements of those related disciplines, provided that these constructs possess (more or less) the same meaning in both domains of inquiry. Unfortunately, various theoretical difficulties confronting these other disciplines are thereby subject to inheritance by sociobiology. Not the least of the problems involved in discovering lawlike relations within molecular genetics, for example, are those arising from epistatic, polygenic and pleiotropic effects, where the respective contributions of various genes in bringing about phenotypical traits tends to be concealed, either because of some non-additive interaction, because more than one gene may contribute to producing certain traits, or because some genes have multiple manifestations. To the extent that sociobiological generalizations are dependent upon discoveries in other disciplines, such as genetics, therefore, these difficulties are implicitly transmitted to sociobiology, although the problems involved in discovering generalizations in sociobiology, admittedly, need not be transmitted back.

This asymmetrical dependence may (at least, in part) explain the vocal tendency of some members of the scientific community to protest that sociobiology should not be taken seriously, until the underlying generalizations from molecular genetics, evolutionary biology, and so on upon which it theoretically depends have been systematically formulated and thoroughly tested; until then, they imply, sociobiology is not deserving of scientific respect, as though research could not be carried on with profit in one discipline unless it had been completed in those other disciplines upon which it depends. If things were this way, we would know very little indeed, for no one would pretend that research in physics is remotely near completion, even with respect to the most fundamental areas of inquiry, including theories of space and time and of quantum mechanical phenomena. One should as well conclude, therefore, that, by such a standard, research in molecular genetics is likewise undeserving of respect, since it is dependent upon discoveries in biochemistry, which in turn are dependent upon discoveries in statistical and quantum mechanics. So if this is how things stand, then the molecular biologists, the bio-chemists and the rest of us might as well make ourselves comfortable, because it looks as though we are in for a long wait! While scientific theories are always subject to the risk of the transmission of falsity from other theories upon which they depend, this is simply one of the hazards of the game: no one should suppose that science is risk-free.

Similar considerations obtain with respect to those portions of the sociobiological vocabulary that are drawn from psychology, sociology and anthropology; for if there is theoretical inconsistency between the findings of social science and of sociobiology, then (as we have already noted) they cannot both be altogether true. As Alexander Rosenberg has observed, it would be wrong to take for granted that the appropriate vocabulary for sociobiological generalizations must correspond to the current vocabulary of contemporary social science (Rosenberg, 1983). Still, it would be an equally great mistake to conclude that terms from disciplines such as these cannot be correct: the language of age-grading, athletic sports, and so forth, after all, belongs to the vocabulary of contemporary social science, yet seems to be ideally suited for the specification of at least one portion of the vocabulary of sociobiology, namely: "culturgens". Indeed, it is difficult to imagine how sociobiological theorizing ought to proceed if not by employing elements of the vocabulary of anthropology, including the categories of incest inhibition, bond formation, and so forth. If sociobiology can be properly criticized, therefore, then it would appear to be far more to the point to take objection to certain other conceptions that Lumsden and Wilson employ, such as those of "mind", of "free-will", and of "morality", especially, as they occur in *Genes, Mind, and Culture* and in *Promethean Fire*. It would be asking a great deal to presume that problems such as these, which have plagued philosophy for more than 2000 years, ought to be resolved, all at a single stroke, by Lumsden and Wilson's theory, but such are its pretensions.

Inherent problems with the central notions of epigenetic rules, interpreted as predispositions to acquire and to manifest specific behavioral tendencies, which are, in turn, interpreted as dispositions, however, would be more serious. Since so much of what has gone before has been intended to clarify their conceptual character, perhaps the rest that ought to be said concerns a matter mentioned only briefly, in particular: that sociobiology may be formulated as an *indeterministic* theory instead of as a *deterministic* theory. For the definitions and conjectures offered above – including 1A and 1B – are advanced as deterministic theses, wherein the same result invariably occurs, under the same conditions (the same behavioral tendency, BT^1, is always acquired or manifest under the same environmental factors, EF^1, by the same phenotype, P), which we formalized by employing the causal conditional of universal strength:

$$\text{(VII)} \quad (x)(t)[Pxt \ni\!\!- (EF^1xt \ni\!\!-_u BT^1xt^*)];$$

yet the same variables can be related by probabilistic causal connections that permit one or another result within a fixed range of possible results to occur, under the same conditions, with different probabilities (and various behavioral tendencies, BT^1, BT^2, and so on, can then be acquired or be manifest under the same environmental factors, EF^1, by the same phenotype, P). Moreover, this too can be formalized by employing a causal conditional of probabilistic strength:

$$\text{(VIII)} \quad (x)(t)[Pxt \ni\!\!- (EF^1xt \ni\!\!-_n BT^1xt^*)];$$
$$(x)(t)[Pxt \ni\!\!- (EF^1xt \ni\!\!-_m BT^2xt^*)];$$

and so forth, which assert, respectively, "If x were of phenotype P at t, then x's being under the influence of environmental factors EF^1 at t would (probabilistically) bring about x's acquiring and manifesting specific behavioral tendency BT^1 at t^*, with strength n", "If x were of phenotype P at t, then x's being under the influence of environmental factors EF^1 at t would (probabilistically) bring about x's acquiring and manifesting specific behavioral tendency BT^2 at t^*, with strength m", etc., where these "degrees of strength" satisfy probabilistic relations of addition, summation, and multiplication (Fetzer, 1981, Part I).

Indeed, this conception is reinforced by Lumsden and Wilson's explanation of behavioral tendencies as properties of individual organisms: their notion of "selectivity" seems to correspond to the size of an outcome space (or of a "sample space") with respect to the variety of different kinds of mentalistic or behavioral reactions R^1, R^2, ... that are available to that organism as potential "modes of response" to its (internal and external) environment EF^i; analogously, their notion of "penetrance" seems to represent the strength of tendency n for that organism to respond with one among those available modes. Moreover, Lumsden and Wilson appear to be very sensitive to the consideration that different neurological and physiological capacities and tendencies characterize particular stages in the growth and maturation of different species, where these changes across time not only affect (the strength of) usage bias curves but also affect (the strength of) epigenetic rules themselves, since, as we are all aware, the capacity for an organism to acquire and to manifest different behavioral tendencies as a function of (physical and social) environmental factors should not be treated as invariable: the changes that occur between childhood, adolescence, maturity, and old age influence not only (the strength of) our dispositions to manifest various behavioral tendencies but (the strength of) our predispositions to acquire them as well. Thus, it would not be unreasonable to anticipate the future elaboration of sociobiology as a broadly indeterministic conception of organisms that develop across time from their genetic origins into cultural creatures influencing the reproduction and survival of the species under the influence of their environments.

Whether sociobiological conjectures are formulated as deterministic or as indeterministic theses, however, they remain empirically testable claims. If epigenetic rules are taken to be universal strength causal tendencies, for example, "theorems" of form (VII) may be tested by attempting to ascertain whether any instance of phenotype P^1 can be under the influence of environmental factors EF^1 and yet fail to acquire or to manifest the behavioral tendency BT^1. Insofar as a set of lawlike hypotheses, such as (VI), logically implies a corresponding set of material generalizations, such as (II), a lawlike hypothesis will be true only if every known instance of P^1 that has been under the influence of EF^1 has acquired or manifest BT^1. But that is not enough, since every instances of P^1 that could be under the influence of EF^1 would have to acquire or manifest BT^1 as well. If epigenetic rules are interpreted as probabilistic strength causal tendencies instead, then "theorems" of form (VIII) may likewise be tested by attempting to ascertain whether instances of phenotype P^1 that are under the influence of EF^1 fail to acquire or to manifest behavioral tendencies BT^1, BT^2, and so on, with frequencies appropriate to the strength of their corresponding probabilities n, m, and so forth. For, while it is logically possible for these frequencies to deviate from such generating probabilities, as the number of instances increases, the likelihood of these deviations becomes quite small. Thus, strictly speaking, although deterministic sociobiological conjectures will be falsifiable but not verifiable, whereas indeterministic sociobiological conjectures will be neither verifiable nor falsifiable, theses of either kind nevertheless qualify as empirically testable claims (Fetzer, 1981, Part III).

Evidence that might be enough to confirm a generalization, however, does not necessarily corroborate the corresponding law, which raises a difficult problem. Different versions of sociobiology will be distinguished by their ascription of different sets of culturgens, epigenetic rules and behavioral tendencies to various species, where these conjectures are intended as lawlike claims. To whatever extent already existing societies exemplify the conditions that would be required to subject these hypotheses to empirical test, moreover, there are no special difficulties; but to whatever extent already existing societies do *not* exemplify those conditions, exceptional problems may be encountered based upon ethical aspects of scientific inquiries. Conjectures about incest inhibition, sex-based infanticide, and other behavioral tendencies are subject to corroboration as lawlike claims only insofar as the non-existence of properties or procedures whose presence or absence would make a difference to the acquisition or the manifestation of those practices can be established. But there are obvious prohibitions against creating conditions of child abuse, of sexual slavery, and so on, by means of environmental manipulations that might violate the principle of treating persons as ends and never as means – even when the object of inquiry is scientific knowledge! As a natural science of social behavior, sociobiology confronts the same ethical issues that constrain experiments involving humans in each of the social sciences; yet these limitations upon human experimentation do not lessen its scientific dimension so much as they elevate its moral character.

Ultimately, the difference between the negative significance of successful attempts to refute an hypothesis (as an aspect of deductive methodology) and the positive significance of unsuccessful attempts to refute an hypothesis (as an aspect of inductive methodology) requires recognition (Fetzer, 1981, p. 176). Here, however, it might be appropriate to point out that sociobiology confronts another difficulty with respect to experimentation, which arises by virtue of changeable parameters of laws of nature themselves, namely: that genetic and cultural evolution can operate at different rates, rendering tests of sociobiological hypotheses more historical and comparative and less experimental and manipulative. An absence of moral imperatives, after all, would not alter the generally slow, Darwinian character of genetic evolution as opposed to the potentially rapid, Lamarckian character of cultural evolution. This difference may be displayed by adapting a Minkowski

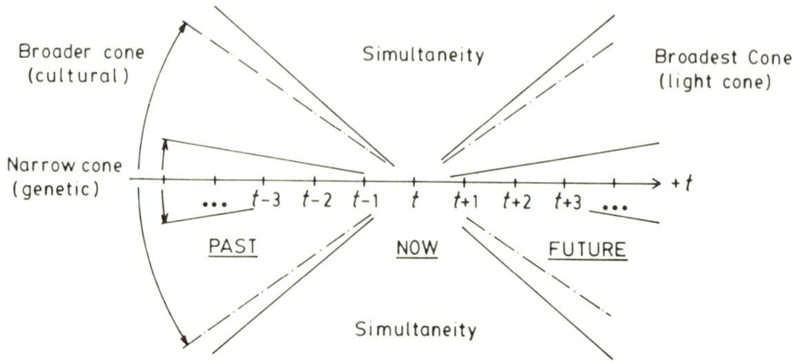

Fig. 1. *Gene-Culture Coevolution Diagram* (Different for distinct species).

space-time diagram to serve as a gene-culture coevolution diagram (Figure 1). where the narrow cone reflects the slower rate of genetic change (as a function of gestation) and the broader cone reflects the faster rate of cultural change (as a function of technology); for, while off-spring can only be produced at a constrained rate of reproduction within a species, the upper bound on the rate of culturgen transmission for an innovative species can approach the speed of light.

One additional distinction that might be drawn with respect to the science of sociobiology was introduced by Sylvain Bromberger (1963), namely: the sense in which the term "theory" may designate a *domain of inquiry* as opposed to that in which "theory" stands for a *set of propositions* pertaining to such a domain. Thus, domains are specifiable by means of questions, while propositions represent instead relatively specific answers. Sociobiology qualifies as a "theory" in both of these senses; yet in our consideration of the scientific standing of this discipline, we have learned a great deal more. Astrology and crystal-ball gazing, after all, are potential "answers" to decision-making "questions", without being "scientific". Those who want to persist in denying that sociobiology is science, of course, might do so by rejecting the intensional construction of lawlike sentences in favor of some extensional alternative. They would then be obligated, not only to provide a means for distinguishing (merely) "accidental" from "lawlike" sentences, presumably, but to acknowledge that sociobiology can still satisfy the standard conception and the semantical conception as well as the Campbellian conception of scientific theories. There appears to be no reasonable criterion of demarcation (concerning its logical structure, its theoretical vocabulary, or its empirical testability) that sociobiology as it has been described above is not able to fulfill. I therefore conclude that sociobiology appears to be as scientific as molecular genetics, evolutionary biology, developmental psychology or cultural anthropology, and suggest that any continuation of "the sociobiology debate" over its scientific significance is more likely to be rooted in the sociology of science than in the logic of scientific procedure.

University of Virginia

NOTE

* A distant ancestor of the present paper was presented at Colorado State University and at the University of Colorado on 23–24 February 1983. I am indebted to David Smillie for his stimulating comments and valuable criticism.

REFERENCES

Braithwaite, R. B.: 1953, *Scientific Explanation*, Cambridge University Press, Cambridge.

Bromberger, S.: 1963, 'A Theory About the Theory of Theory and About the Theory of Theories', in B. Baumrin (ed.), *Philosophy of Science: The Delaware Seminar*, Vol. 2, Interscience, New York, pp. 79–105.

Campbell, N. R.: 1920, *Physics: The Elements*, Cambridge University Press, Cambridge. Republished with the title, *Foundations of Science* (Dover Publications, New York, 1957).

Fetzer, J. H.: 1981, *Scientific Knowledge*, D. Reidel, Dordrecht, Holland.

Giere, R. N.: 1979, *Understanding Scientific Reasoning*, Holt, Rinehart and Winston, New York.

Hempel, C. G.: 1965, *Aspects of Scientific Explanation*, The Free Press, New York.

Hempel, C. G.: 1966, *The Philosophy of Natural Science*, Prentice-Hall, Englewood Cliffs.

Leach, E.: 1981, 'Biology and Social Science: Wedding or Rape?', *Nature* **291**, 267–268.

Lewin, R.: 1982, 'How Did Humans Evolve Big Brains?', *Science* **216**, 840–841.

Lumsden, C. R. and Wilson, E. O.: 1981, *Genes, Mind, and Culture*, Harvard University Press, Cambridge.

Lumsden, C. R. and Wilson, E. O.: 1983, *Promethean Fire*, Harvard University Press, Cambridge.

Popper, K. R.: 1968, *Conjectures and Refutations*, Harper & Row, New York.

Rosenberg, A.: 1983, 'Human Science and Biological Science: Defects and Prospects', in N. Rescher (ed.), *Scientific Explanation and Understanding*, University Press of America, Lanham, pp. 37–51.

Ruse, M.: 1977, 'Is Biology Different From Physics?', in R. Colodny (ed.), *Logic, Laws, & Life*, University of Pittsburgh Press, Pittsburgh, pp. 89–127.

Sneed, J. C.: 1971, *The Logical Structure of Mathematical Physics*, D. Reidel, Dordrecht, Holland.

Suppes, P.: 1967, *Set-Theoretical Structures in Science*, mimeographed. Stanford.

Wilson, E. O.: 1971, *The Insect Societies*, Harvard University Press, Cambridge.

Wilson, E. O.: 1975, *Sociobiology: The New Synthesis*, Harvard University Press, Cambridge.

Wilson, E. O.: 1978, *On Human Nature*, Harvard University Press, Cambridge.

Wilson, E. O.: 1978a, 'Foreword', in A. Caplan (ed.), *The Sociobiology Debate*, Harper & Row, New York, pp. xi–xiv.

Wilson, E. O.: 1980, *Sociobiology: The Abridged Edition*, Harvard University Press, Cambridge.

EPILOGUE

MICHAEL RUSE

EVOLUTIONARY EPISTEMOLOGY: CAN SOCIOBIOLOGY HELP?

ABSTRACT. The sociobiologist Edward O. Wilson argues that humans think and act in ways constrained by certain innate dispositions: epigenetic rules. These have been put in place by evolution through natural selection. Expanding this empirical claim, the thesis of this paper is that epigenetic rules throw significant light on major problems in the theory of knowledge. It is argued also that the precursor to Wilson (in this respect) was David Hume.

Camus said that the only serious philosophical question is suicide. That is wrong even in the strict sense intended. The biologist, who is concerned with questions of physiology and evolutionary history, realizes that self-knowledge is constrained and shaped by the emotional control centers in the hypothalamus and limbic system of the brain. These centers flood our consciousness with all the emotions – hate, love, guilt, fear, and others – that are consulted by ethical philosophers who wish to intuit the standards of good and evil. What, we are then compelled to ask, made the hypothalamus and limbic system? They evolved by natural selection. That simple biological statement must be pursued to explain ethics and ethical philosophers, if not epistemology and epistemologists, at all depths. (Wilson, 1975, p. 1)

Charles Darwin's *On the Origin of Species*, published in 1859, argued that the organic world, including ourselves, evolved by a slow process, which Darwin called "natural selection". More organisms are born than can possibly survive and reproduce, and in the conequent "struggle for existence" certain ("fitter") organisms out-perform the rest. Thus we have a natural process of winnowing or "selection". Given enough time, this leads to full-blooded evolution and to the development of distinctive organic features which help their possessors in the struggle as "adaptations". (Full details in Ruse, 1979a; 1982a)

These evolutionary ideas did not in themselves spell the end of conventional religious belief. But they did severely jolt people's confidence that a Wise Father stands behind the universe, sustaining His work, and guaranteeing that His most favoured creation – man – has privileged insight into the true workings of reality. It is, therefore, not surprising to find that after the *Origin* many people turned to evolution itself, searching for a new foundation to human knowledge. If Christianity cannot give us truth, then perhaps the new science can. Thus, there grew up – and has endured – a school of thought that attempts to find, within evolutionary thinking, the keys to the growth,

James H. Fetzer (ed.), Sociobiology and Epistemology, 249–265.

development, and maintainance of human understanding (Campbell, 1974; Plotkin, 1982).

Generally, advocates of such an "evolutionary epistemology" have grasped at what they believe to be a firm analogy between the evolution of organisms and the growth of human knowledge, particularly scientific knowledge. Hence, it has been argued that the probable (and only) cause for the success of knowledge lies in some mechanism analogous to natural selection. It is claimed that just as organisms must prove their worth in the ongoing struggle for existance, so also ideas must prove their worth against other ideas in an intellectual "struggle for existence". And, argue supporters of this analogy, this is all one can have or hope for in the realm of knowledge. The best science somehow works better than its rivals. (See, for instance, Toulmin, 1972; Popper, 1972.)

In the past century, this position has been adopted by many leading scientists and philosophers, including (then) Thomas Henry Huxley and (now) Sir Karl Popper. However, as with all analogies, there are points where these two fields – organic evolution and epistemological development – do not match, and in my opinion (as well as that of many others) the points of mismatch are so great as to render the whole enterprise very suspect indeed (Ruse, 1983). "Evolutionary epistemology" was a nice idea, but it simply will not work.

Recently, however, it has been argued that perhaps the case is not entirely closed. Most particularly, the Harvard biologist Edward O. Wilson insists – stridently – that Darwinian evolutionary theory is relevant to our understanding of knowledge. He indeed claims that biology has things to say of importance to the most basic philosophical epistemological concerns. Philosophers ignore evolutionary theory at their own peril. In particular, argues Wilson, philosophers ignore at their peril the extension of neo-Darwinian thought to the study of the biological basis of human social behaviour – so-called human "sociobiology".

I have now come to the decision that Wilson is right. Epistemology needs Darwinism. To show this, I must first spell out Wilson's pertinent scientific and philosophical claims. In part because I have dealt with such questions in depth elsewhere, and in part simply because I want to get the discussion moving, I shall assume in this essay that, qua science, these claims of Wilson are well-taken. Once they are laid out, we can explore their philosophical implications. (For more on sociobiology, see Caplan, 1978; Ruse, 1979b, 1982a. Wilson's views on humankind are to be found in Wilson, 1975, 1978; Lumsden and Wilson, 1981, 1982.)

ON HUMAN NATURE

Human beings evolved from ape-like creatures in the past 5 million years or so. Our own species, *Homo sapiens*, has been around for the last 500,000 years. As with other organisms, the chief causal factor in human evolution was natural selection. The exact nature of this selection, and how it affected such distinctively human features as our brain and our hands, is still debated; but, that it was crucially important is not a matter of real dispute. Nevertheless, today, we humans are *prima facie* very different from other denizens of the organic world. In particular, we have our culture. We have language, and artifacts, and customs, and morality, and religion, and much, much more. Animals seem at most to have only traces of all of the things which we associate with being human. In a sense, therefore, we seem to have escaped our biology.

Wilson argues that, although by virtue of their culture, humans are indisputably different from other animals, to think that we are entirely different or that we have escaped our biology is totally to distort the true picture. Indeed, beneath our culture, structuring it and pushing through in many places, we are a typically mammalian species. Most importantly, our much-vaunted sense of morality is simply the biological product of our need to get along with our fellow species members, and is something to be found in all the higher animals. At least, it is to be found in all the higher animals which (like humans) lead a social existence. And even our power of reason is little more than another tool in the battle for biological supremacy.

> The essence of the argument, then, is that the brain exists because it promotes the survival and multiplication of the genes that direct its assembly. The human mind is a device for survival and reproduction, and reason is just one of its various techniques. (Wilson 1978, p. 2)

Where lies culture – which presumably includes all of the knowledge claims we would want to make – in the true picture of humankind? For Wilson, through the power of reason, it is yet one more adaptation (or complex of adaptations) making *Homo sapiens* the highly successful species that it is – or rather, making individual members of the species the highly successful organisms that they are. In short, culture must be understood as a phenomenon kept in firm check by the ultimate units of biological control: "The genes hold culture on a leash." (Wilson, 1978, p. 167)

In this essay, Wilson's claims are being taken as scientifically sound. But even if one grants this much, as far as our own *philosophical* inquiry is concerned, not much progress seems to have been achieved. We can grant that, in some broad sense, knowledge (from now on, for simplicity, let us restrict the discussion to scientific knowledge) has adaptive value. The person, or the society, who has a grasp of the principles of (say) mechanics is probably going to be better able to exploit the environment than the person or society who does not have such a grasp. But, this is all very broad and general, virtually to the point of being a truism. Biology seems to give little clue to the actual content and nature of science. No one thinks the explorer lost in the jungle will be better off for know that $F = ma$. Nor does anyone think that the person with a better scientific theory is going to be that much more of a reproductive success. After all, Copernicus, Descartes, and Newton – to name three of the giants of the Scientific Revolution – died childless!

This is all true. However, in his more recent writings – specifically, those co-authored with Charles Lumsden – Wilson puts forward ideas which speak to these very concerns. Now, he is in a position to show very much more precisely just how it is that biology can enter right into the content of science. Biology can address philosophical questions.[1]

EPIGENETIC RULES

The central notion in Wilson's new theory – or, more accurately, Wilson's new extension to his earlier thought – is that of a "culturgen". This is a basic unit of culture, and can be one of a number of different things. an artifact, a specified piece of behaviour, a concept (a "mentifact"), or any element one would identify as being part of a person's or a society's general cultural dimension. Although Wilson does not pick out science for particular attention, presumably a scientific theory is comprised of culturgens, and any specific theory can be compared and distinguished from any other theory in terms of similarities and differences in culturgens.[2]

The problem which Wilson now faces is how the culturgens got learnt, organized, altered, passed from one generation to the next, and how these processes relate back to the genes. Wilson explains such processes and connections through the concept of an "epigenetic rule". These are biological constraints on development and on our consequent capacities for learning, Wilson (and Lumsden) divide the epigenetic rules into two sets. First, there are those which affect and organize the reception of the basic items of information about the world and ourselves, and so forth. Included here are colour sensations, sounds and smells, and the like. Second, there are those rules which organize this basic information, enabling us to act in various ways upon it. Thus, information comes *via* the primary rules, and this is then organized, structured and evaluated *via* the secondary rules.

How would this all work in the case of science? On the one hand, science is empirical. It is based on the evidence of the senses. It is nothing if not that. Here, presumably, we are getting information as filtered through the primary rules. On the other hand, science is not just a random collection of empirical notions. It is a highly formalized enterprise with a definite methodology, and with canons dictating what is acceptable and what is non-acceptable. Here the secondary rules come into play.

What sorts of this would be produced through the influence of the secondary rules? Most obvious would be basic mathematics and logic. Science is built on the principles that it is true that "2 + 2 = 4" and that it is false that "2 + 2 = 5". Science also accepts that if a pressure is 5 lbs / square inch, it cannot simultaneously be 10 lbs / square inch. Note that this is not to bar the use of alternative mathematics or of alternative logics, although one would not use rivals at the same time in the same place.

There is more to the methodology of science. A much debated topic is that of exactly how crucial laws are to the scientific enterprise. Some think all explanation demands reference to law. Others are not so sure (Hempel, 1966; Goudge, 1961). However, all agree that thinking in terms of laws is central to science. Even more central is thought in terms of *causal* laws. Forces cause bodies to fall. Continental drift causes changes in the geographical distributions of organisms. Clearly then, one epigenetic rule (or set of such rules) directs us to think causally. Given the same initial conditions, we expect the same consequences – and if we do not get them, we do not throw out our causal thinking, but rather we persist until we find some way of bringing our experience in line with our expectations. No one thinks that the boomerang violates Newton's laws, nor did anyone think so when they first saw one thrown. Along with the drive for causal laws, within science we find a push towards simplicity and unification. The best science, whether it be physics, biology, geology, or whatever, aims to show how different ideas and areas all bind together into one unified whole. Once one accepts that the continents do move, one has an explanation of earthquakes, of volcanoes, of continental shapes, of organic distributions, of deep-sea rifts, and much, much more. Everything fits simply together, in what the nineteenth century philosopher William Whewell termed a "consilience of inductions" (Whewell, 1840; Kitcher, 1981; Ruse, 1982b).

Why do we value a consilience? Because, as with logic, mathematics and causality, epigenetic rules make us think that the best science is consilient.

And so the story goes. But we need read no further, for now we can grasp the outlines of a Wilsonian analysis of science. There is certainly no claim that every last item in a scientific theory has direct biological value. There is not even the claim that a theory judged better by scientists will confer superior biological power on its supporters. Therefore, one should not look for an immediate jump in the reproductive success of the Einsteinian over the Newtonian, or the Evolutionist over the Creationist.

However, viewed through the lens of sociobiology, there is nevertheless a very strong connection between science and evolutionary advantage. Moreover, this is an advantage meant to be understood in a literal not a metaphorical sense. The mathematics of science, the logic of science, the methodology of science, emerge as products of the secondary epigenetic rules, and these rules have supposedly been forged in the evolutionary struggle for survival and reproduction. Thus biology informs and dictates the very skeleton of science.

This means, of course, that even though one scientific theory *per se* may not confer reproductive value over another, choice between scientific theories is governed by biology. Therefore, in the long run one expects to see (scientifically judged) superior science prove of reproductive value to its holders. And this is so. Consider, for instance, why a scientist prizes predictive fertility in a scientific theory. Why is it thought to be a merit of a scientific theory that we can draw checkable inferences from it, and thus why is it (say) thought that a great merit of continental drift theory was its ability to predict great similarities in the geological patterns of Africa and South America? Why does one care whether a theory can do this or not, and why should this be a major factor in one's choice of one theory over another?

The answer simply is that those proto-humans who took predictive fertility seriously, tended to out-survive and out-produce those that did not. The Australopithecine who (say) took note of changing leaf-colour and who then predicted coming climactic changes (and took appropriate action) tended to get through the winter a lot better than his rival, who blithely frittered away the summer and fall.

Again, take the notion of a consilience. The pre-human who refused to take coincidences at face-value, but who looked for underlying connecting causes, was at an adaptive advantage over the credulous. Sounds in the jungle, marks at the water-hole, traces of blood, add up to tigers – even though they may be unseen. Beware!

Today, in our science, this kind of inference is all very much sanitized. The average physicist is about as far from a cave-man as it is possible to imagine. Refusal to accept unseen entities may lose you the Nobel Prize. It tends not to make you an animal's dinner. But, cave-man and physicist are linked by the same epigenetic rules. Sophisticated modern science is simply jungle-lore writ fine and small. Moreover, not only are the reasons why we accept one modern scientific theory rather than another – and why indeed we create and accept science at all – firmly rooted in our biology, but even now adaptive advantage is not totally divorced from the success of science. As modern science has gotten more sophisticated, and thanks to this very sophistication, its creators have way outstripped their previous reproductive efforts. The methods of science, applied in technology and medicine, are superb adaptations for the struggle for reproduction.

In short, today as always the genes hold culture on a leash (Wilson, 1978, p. 167).

THE NATURE OF SCIENCE

Thus far I have been trying to spell out the way in which a Wilsonian approach would treat science: how it is formed and its criteria which might lead to change. In order to do this, I have drawn on some of the standard results of the philosophical analysis of science, for instance about the importance of causal understanding and about the drive to consilience. To deepen our understanding of this neo-Darwinian epistemology, let us now turn the tables and ask about the kind of philosophical understanding of science towards which Wilson points.

This may sound somewhat circular, and in a sense it is – but not viciously so. I am not suggesting that, for instance, the philosopher's identification of the importance of simplicity is the sole reason for accepting an epigenetic rule centering on simplicity, which then becomes the only reason for arguing that simplicity is important in science. Rather, philosophy (along with other disciplines) can identify certain things that scientists seem to take seriously. The sociobiologist like Wilson can then claim that they are at the centre of epigenetic rules. Of course, independent empirical evidence must be given for such claims. For the sake of argument, in this essay, we are assuming that such evidence is forthcoming. But then, given the rules, we may legitimately ask about the nature of the end product of thought, namely the finished science. This is what I want to do now.

Philosophers divide over the ultimate nature of science and the reasons for acceptance of one theory rather than another. Until recently, orthodox opinion had it that it is the empirical evidence that is decisive. Does a particular theory "fit the facts"? If it does, then this is good reason to accept the theory. If it does not, then we must go on looking. One well-known version of this viewpoint is that of Sir Karl Popper (1959). A genuine scientific theory must be falsifiable, and it is this measure against the real world that is absolutely decisive. However attractive a theory may seem, if it goes against the facts it must be rejected.

In the past two decades, thanks particularly to the work of such thinkers as Thomas Kuhn (1962), many have started to emphasize strongly the inadequacy of any such straight-forward appeal to "the facts".[3] One has to (and does in fact) invoke other criteria in theory-choice – criteria like simplicity and elegance. Some thinkers, primarily sociologists and historians, have been so impressed by these non-evidential criteria, that they have downgraded the empirical side of science to little more than a shadow. Ideology and like factors are virtually everything – they are certainly virtually everything which matters.

Wilson's analysis helps one to tread a careful middle-line through this debate – a middle-line which has the virtue of staying true to the way in which scientists actually behave. On the one hand, the Wilsonian agrees that the "facts" are absolutely basic determinants in the nature and progress of science. Science is nothing if it is not an attempt to come to terms with reality. One is trying to understand the world through abilities which enable one better to survive and reproduce in the world. (This is not to say that the "facts" of science are ever absolutely uninterpreted. Indeed, the primary epigenetic rules probably ensure that they never are.)

On the other hand, the Wilsonian expects there will be more at stake than just the facts. For instance, thanks to natural selection, notions of simplicity are going to be crucial, in science as elsewhere in life. One chooses the simplest curve to link the dots, not because it has some ontological claim on being right – or because it is absolutely determined by the facts – but because it is the simplest! It is the curve which requires least human hassle, as we strive to manage and use the information of our senses. To put the matter bluntly, the cave-man with a taste for the complex tended still to be working out the mathematics, whilst his simple-minded cousin was getting right down to feasting and copulating.[4]

In short, the Wilsonian sees science very much from a pragmatic perspective.[5] In a sense, therefore, there is a close affinity between a Wilsonian position and the naturalistic approach to philosophy advocated by W. V. Quine. Parenthetically, it is interesting to note that Quine himself has remarked on the Darwinian nature of his views. In one well-known passage, he writes:

> One part of the problem of induction, the part that asks why there should be regularities in nature at all, can, I think, be dismissed. *That* there are or have been regularities, for whatever reason, is an established fact of science: and we cannot ask better than that. *Why* there have been regularities is an obscure question, for it is hard to see what would count as an answer. What does make clear sense is this other part of the problem of induction: why does our innate subjective spacing of qualities accord so well with the functionally relevant groupings in nature as to make our inductions tend to come out right? Why should our subjective spacing of qualities have a special purchase on nature and a lien on the future?
>
> There is some encouragement in Darwin. If people's innate spacing of qualities is a gene-linked trait, then the spacing that has made for the most successful inductions will have tended to predominate through natural selection. Creatures inveterately wrong in their inductions have a pathetic but praise-worthy tendency to die before reproducing their kind. (Quine, 1969, p. 126.)

I would suggest that Wilson's science enables us to put empirical flesh on Quine's skeletal philosophical suggestions.[6]

But, suppose we grant the essential truth of the picture of science being sketched here. Suppose we agree that scientific knowledge is regulated according to certain principles or norms which have their foundation in our evolutionary heritage. At once a pressing question intrudes. It is obvious that even in the most advanced societies – societies which have pushed science to its highest peaks – many people refuse to take science as seriously as a Wilsonian analysis would seem to merit – and demand. Astrology flourishes. Varieties of pseudo-medicine abound. And, as the recent controversy over Creationism well-attests, the crudest forms of religious belief soak up enthusiasts and sympathizers (Futuyma, 1983; Kitcher, 1983).

A more conventional analysis of science – one which supposes that there is an objective reality and that science attempts to map or capture it – presumably explains such alternatives to science in terms of error, and leaves matters at that. People confuse the true nature of causal understanding, or some such thing, and that is all there is to it. We all make mistakes. However, the kind of position we are considering cannot dismiss such rival ways of thinking quite so readily. Science is supposed to be based on rules which confer survival value. How then can so many people ignore or modify the rules? What possible survival value could there be in the liberties in causal thinking that such phenomena as astrology or Creationism demand?

In fact, what may seem at first to be a refutation of our neo-Darwinian evolutionary epistemology turns out to be strong support. To the sociobiologist like Wilson such scientific alternatives are virtually to be expected. Being human requires compromises. We are not simply calculating machines, trying to understand and master nature, with no other goals. We are also social beings, with needs in that direction – with needs, for instance, to get on with our neighbours. This requires ethics, and if (as it happens) religion can help to reinforce and maintain ethics, then this is a good selective reason why people should have a biological disposition towards religious (and like) beliefs, even if they may clash with science.

For another major reason why people embrace notions of supernatural beings, able to break at will with the regular course of nature, one need look no further than the Old Testament and the story of Job. People need crutches like religion to help them to get through life – to give meaning to their daily existence and to help and comfort in times of stress. Why did not Job curse God, which certainly would have been the rational thing to do? Because, as the astute psychologist who wrote the story saw full well, it is precisely when things go wrong that you need something to help you to get through: to go on living (literally!). And if that something is belief in an entity (or system) which can break or soften the harsh realities, and which can give meaning and purpose too, then so be it.

I do not think it is necessary to push the point any further. Science is one part, a very important part, of being human. But there are other human needs, and these mean that, as a matter of fact, most people will simply not be ruled by science all of the time. Indeed, in a sense, the sociobiologist like Wilson is somewhat of a freak. He takes the basic premises of science and pushes them through to explain the rest of the human dimension. Most people genuflect towards science (even the Creationists pay lipservice), but then calmly break from it when it suits them. Darwinian evolutionary epistemology shows why.

PHILOSOPHICAL PREDECESSORS: KANT?

Like most scientists, Wilson tends to be somewhat ahistorical. He writes as though his ideas represent a total break with anything from the past, and as if no one – certainly no philosophers – has ever thought in ways parallel to his. However, even the most innovative of positions usually turns out to be not quite that much of a leap into new territory, and I rather suspect that this is true of Wilsonian implications for philosophy. At least, this is what I shall now argue. I do not try to link up Wilson's views with those of other thinkers in order to belittle his importance. Rather, I hope thereby to throw deeper light on sociobiologically inspired evolutionary epistemology.

What is being argued is that our knowledge of the external world is not something just 'read off", as it were, from experience, as crude empiricists are supposed to claim. Rather, such knowledge is a product of the mind actively interfering with and interpreting the date of the senses. The epigenetic rules filter and structure experience. Claims like this at once put one in mind of the greatest of all modern philosophers. Immaneul Kant, for he also argued that empirical knowledge is not passively received by the mind, but is rather something made and structured by the mind acting on the evidence of the senses.[7] In other words, our empirical "synthetic" knowledge is infused by elements which do not come from the external world and which are thus "*a priori*":

> Hitherto it has been assumed that all our knowledge must conform to objects. But all attempts to extend our knowledge of objects by establishing something in regard to them *a priori*, by means of concepts, have, on this assumption, ended in failure. We must therefore make trail whether we may not have more success in the tasks of metaphysics, if we suppose that objects must conform to our knowledge. (Kant, 1929, p. 22.)

Kant's success in exploiting this insight was his so-called "Copernician revolution" in philosophy, seeing that we actively create knowledge rather than passively receiving it.

The case for regarding Wilson as thinking in a neo-Kantian fashion is strengthened when one starts to consider some of the specific claims of their respective positions. Thus, for instance, Kant argues that claims of mathematics like "5 + 7 = 12" are claims based on experience but interpreted by the mind. They are "synthetic *a priori*" claims. Similarly, Kant argues that causal claims like "the lead ball falling on the cushion caused the depression" are mind structured. "We can extract clear concepts of them, i.e. causes from experience, only because we have put them into experience, and because experience is thus itself brought about only by their means." (Kant, 1929, p. 223)

We have seen that both mathematics and areas requiring causal understanding are subjects where Wilsonian epigenetic rules would play crucial roles. The human who thought that "5 + 7 = 12" rather than "5 + 7 = 13" would be fitter than one who thought otherwise. The human who realized that falling lead balls are liable to cause depressions – in cushions, heads, or whatever – would likewise be fitter than one who thought otherwise. Thus, in both cases, the human mind has been fashioned by evolution to see the world in the manner suggested by Kant.

But, even if we agree on the mind-given element, there is more to mathematics and causality than this. And here, I begin to wonder just how deeply

Wilson is entrenched in a Kantian framework. The point about mathematics and causality and the like claims is that we don't simply believe them, but that we believe them to be *necessary*. We think that 5 + 7 *must* equal 12, and that, given the falling ball striking the cushion, there *must* be a depression. Now, both Kant and Wilson acknowledge this necessity, but I sense radically different attitudes underlying these acknowledgements.

For Wilson, the necessities of knowledge are all very much contingent, psychological matters. It so happens that if you shove your hand in the fire, damage is done to the physical body. Thus there was adaptive advantage in setting up pain sensations, a belief that fire "causes" pain, and a corresponding fear. But, if the world had been different, then so be it. No necessary connections would have been drawn. Similarly, in mathematics and other subjects. Necessity really is in the eye of the beholder, and is a quirk of accident and selection. If there had been some other, more efficient way for selection to the job, we might not have had any sense of necessity at all. After all, we think ourselves free when it comes to morality!

For Kant, things are different. There is nothing accidental about his necessity. Even if it starts in the mind, the necessity of the synthetic *a priori* comes as a condition of the possibility of any rational though whatsoever. Having allowed the importance of sensation. Kant asks rhetorically:

> The question now arises whether *a priori* concepts like causality do not also serve as antecedent conditions under which alone anything can be, if not intuited, yet thought as object in general. In that case all empirical knowledge of objects would necessarily conform to such concepts, because only as thus presupposing them is anything possible as *object of experience* (Kant, 1929, p. 126, his italics.)

And Kant's answer, resoundingly, is that such necessary conformation is a feature of empirical knowledge.

In short, although Wilson's approach is Kantian in that he invokes a mind-structuring element in knowledge, in spirit it seems very different. For Kant, the *a priori* has a hold on knowledge which seems lacking from the form that Wilson's epigenetic rules impose on knowledge. For the evolutionist, in a very real sense, had things been otherwise, one could imagine different rules and thus differently structured knowledge.

Parenthetically, having earlier been critical of Popper, let me freely note that (his other evolutionary epistemological speculations apart) he has seen the Kantian-like implications of Darwinism for humans, and also he has seen that Kant proves too much for a Darwinian.

> In trying to show how knowledge is possible, he proposed a theory which had the

unavoidable consequence that our quest for knowledge must necessarily succeed, which is clearly mistaken. When Kant said, "Our intellect does not draw its laws from nature but imposes its laws upon nature", he was right. But in thinking that these laws are necessarily true, or that we necessarily succeed in imposing them upon nature, he was wrong. (Popper 1962, p. 48.)

PHILOSOPHICAL PREDECESSORS: HUME?

This means that, despite important overlap between Kant and Wilson, if we are to find philosophical antecedents to Wilsonian epistemology, we must dig more deeply. And, for a number of reasons, one possible precursor stands out preeminently, namely David Hume.

First, Kant himself was responding to problems set by Hume, so it could well be that Wilson's *prima facie* similarity to Kant is a function of a similar response to the same issues. Second, there are historical reasons to look towards Hume, since Wilson is working directly in the Darwinian tradition, and Darwin in turn was much influenced by British empiricism. Third, unlike Kant, Hume was always stressing the links between human thought and animal thought, the very basis of Wilson's sociobiological programme. Kant, like the rationalists, had a low view of the status of animals. He did not think them conscious beings or in any way capable of what we would call reasoned thought. They are basically machines, "*Apprehensio bruta* without consciousness" (Smith, 1923, p. x/ix, quoting a letter from Kant to Furst von Beloselsky, 1792). Hume wrote: "(N)o truth appears to me more evident, than that beasts are endow'd with thought and reason as well as men" (Hume, 1978, p. 176).

Fourth and finally, there is a deciding factor which lies outside the central theme of this essay: in ethics, Wilson's position is simply a development of that of Hume (Murphy, 1982; Ruse, 1984). Unlike Kant, both men see ethics rooted in human emotions or sentiments, and think that ultimately ethics must be related back to the individual, and his/her aims and desires. (This does *not* mean that either push some crude ethical egoism.) Given that for both Hume and Wilson their moral theories are part of their overall philosophies – indeed, in both cases theories of morality are probably the inspirations for epistemologies – the case for looking to Hume as the true philosophical ancestor of Wilson becomes overwhelming.

Expectations are fulfilled! When one turns to Hume's work, particularly his *Treatise*, Wilsonian epistemology can readily be seen as progress down the path first trodden by the great Scottish philosopher. Consider Hume's central discussion, on which so much so much of his theory depends: his deservedly celebrated analysis of causal connection. Hume argues that we do not see any necessity in nature. We see only that one event is being followed by another. Nevertheless, we impute necessity to nature. We pick out phenomena which we identify as "causes", and we believe that they must be followed by phenomena we identify as "effects". We say that the fire causes pain and burning, and that the effect (pain and burning) necessarily follows on the cause (fire). Moreover, we distinguish such cases from merely accidental successions. No one thinks there is any necessity between (say) a rainstorm and the new moon then arising.

Hume argues that the mind, as it were, reads necessity into nature. What happens is that we see that those things which we call "causes" are constantly followed by those things which we call "effects", and so this sets up a notion of necessity in the mind, which then puts it into nature, as though it were an external power. " 'Tis the constant conjunction of objects, along with the determination of the mind, which constitutes a physical necessity" (Hume, 1978, p. 171). But why do we think necessity is objective?

> 'Tis a common observation, that the mind has a great propensity to spread itself on external objects, and to conjoin with then any internal impressions, which they occasion, and which always make their appearance at the same time that these objects discover themselves to the senses. (Hume, 1978, p. 167.)

Let us spell out what's going on here. Events follow each other with regularity in nature. Humans notice these successions, and they trigger in the mind expectations and feelings of necessity. Why is there such a triggering? Because of the way the mind is made. It has a propensity to see necessity in successions themselves. However, the content of the successions is given by nature. Then, finally, the mind (unbeknownst to itself) reads necessity into nature, thinking nevertheless that it has found it as an objective facet of nature (Wolff, 1960).

This is Wilson's position almost literally. The only thing missing is the terminology of "epigenetic rules". These are Hume's "propensities". This is not to say that Wilson would endorese all of Hume's psychology, for instance about the way in which the propensities/epigenetic rules would be triggered into action. Nor is it to deny Wilson's advance in seeing that the reason the mind works as it does lies in evolution. But it is to say that Wilson is carrying on David Hume's programme, not breaking out entirely anew. Wilsonian evolutionary epistemology is Hume's philosophy viewed through the lens of neo-Darwinian sociobiology. Thus, new science links harmoniously with old philosophy.

CONCLUSION

There are good reasons for treating with suspicion any new-world philosophy: a supposedly totally fresh way of looking at problems of knowledge. Too often, such claims turn out to be flawed rehashes of long explored ideas from the past. But, this is not to say that advance in philosophy is impossible. It has happened before, and we might hope that it will happen again. My suspicion is that an evolutionary epistemology of the type sketched in this essay may be one such advance.

I confess that I am embarrassed by grandiose claims of this kind. Whether my suspicion is well-founded must be decided by my readers. Let me simply note two pointers which suggest that this approach to epistemology may not be doomed to the graveyard of overblown metaphysics. First, history shows well that advances in philosophy are often triggered by (and based on) advances in science. One has only to think of the effect of the Scientific Revolution. Both Hume and Kant were influenced by Newton. Could it be that the time has come for the Darwinian Revolution to make its mark on philosophy? Second, history shows equally well that advances in philosophy always build on the past. One takes up the torch held high by the great thinkers and carries it further into the darkness. I have argued that neo-Darwinian evolutionary epistemology builds on the thinking of the giants of the eighteenth century – Kant and (even more) Hume. This is no guarantee of its ultimate validity, but (as with the first pointer) it is a strong reason for taking it seriously.

ACKNOWLEDGEMENT

This essay was written during a time of support by the John Simon Guggenheim Memorial Foundation.

University of Guelph, Ontario

NOTES

1 I trust I am not doing Lumsden an injustice by thinking of the ideas to be expounded (drawn primarily from *Genes, Mind and Culture*) as essentially Wilsonian. If I am, I apologize.

2 Wilson is concerned to lay out the basic position. In both exposition and discussion. I have freely exteded his ideas to the problem of scientific knowledge.

3 Professional philosophers of science knew full well before Kuhn that there is more to

science than just the empirical evidence. But, it is not so much of a parody to say that people like Popper do emphasize the appeal to the world of experience, and people like Kuhn deemphasize such an appeal. If this were a paper on the philosophy of science. I would certainly have to qualify my broad claims.

4 I draw your attention to an excellent discussion of this very point by Elliot Sober (1981), who shows that some of the more innane topics to have obsessed philosophers of science in recent years are easily dissolved once one takes evolution seriously.

5 Is Wilson's philosophy of science "Pragmatic" as opposed to "pragmatic" (*i.e.*, does Wilson fall readily into the Pragmatic school of philosophy)? The short answer is that in important respects, it is (and he does). Like the Pragmatists, the Wilsonian mind contributes crucially to knowledge and he relates truth to human nature. But, although the Pragmatists were certainly sympathetic to biological evolution, they differed strongly from Wilson on the reasons why humans interpret the world as they do. The Pragmatists thought culture, not biology, is the key factor. Thus, William James:

> My thesis now is this, that our *fundamental ways of thinking about things are discoveries of exceedingly remote ancestors, which have been able to preserve themselves throughout the experience of all subsequent time*. They form one great stage of equilibrium in the human mind's development, the stage of *common sense*. (James, 1907, p. 170, his italics.)

6 See Lewis (1979) and Cooper (1979) for discussion of Quine's Darwinian views.

7 Wilson's pragmatism is in itself enough to make one think of Kant. C. S. Peirce, the first and deepest of the Pragmatists, was much influenced by the great German philosopher.

REFERENCES

Campbell, D. T.: 1974, 'Evolutionary Epistemology', in P. A. Schilpp (ed.), *The Philosophy of Karl Popper*, Open Court, La Salle, pp. 413–63.

Caplan, A.: 1978, *The Sociobiology Debate*, Harper & Row, New York.

Cooper, D. F.: 1979, 'The Argument From Evolution', *The Aristotelian Society Supplementary Volume LIII*, Compton Press, London, pp. 223–37.

Darwin, C.: 1859, *On The Origin of Species*, John Murray, London.

Futuyma, D.: 1983, *Science on Trial*, Pantheon, New York.

Goudge, T. A.: 1961, *The Ascent of Life*, University of Toronto Press, Toronto.

Hempel, C. G.: 1966, *Philosophy of Natural Science*, Prentice-Hall, Englewood Cliffs.

Hume, D.: 1978, *A Treatise of Human Nature*, Clarendon Press, Oxford.

James, W.: 1907, *Pragmatism: A New Name for Some Old Ways of Thinking*, Longmans, Green, New York.

Kant, I.: 1929, *Critique of Pure Reason*, trans. N. K. Smith, MacMillan, London.

Kitcher, P.: 1981, Explanatory Unification, *Phil. Sci.* 48, 507–531.

Kitcher, P.: 1983, *Abusing Science*, MIT Press, Cambridge, MA.

Kuhn, T. S.: 1962, *The Structure of Scientific Revolutions*, University of Chicago Press, Chicago.

Lewis, H. A.: 1979, The Argument From Evolution, *The Aristotelian Society Supplementary Volume LIII*, Compton Press, London, pp. 207–21.

Lumsden, C. J. and Wilson, E. O.: 1981, *Genes, Mind and Culture*, Harvard University Press, Cambridge, MA.
Lumsden, C. J. and Wilson, E. O.: 1983, *Promethean Fire*, Harvard University Press, Cambridge, MA.
Murphy, J. G.: 1982, *Evolution, Morality, and the Meaning of Life*, Rowman and Littlefield, Totowa, NJ.
Plotkin, H. C.: 1982, *Learning, Development, and Culture: Essays in Evolutionary Epistemology*, John Wiley, Chichester.
Popper, K. R. 1959, *The Logic of Scientific Discovery*, Hutchinson, London.
Popper, K. R.: 1962, *Conjectures and Refutations*, Routledge and Kegan Paul, London.
Popper, K. R.: 1972, *Objective Knowledge: An Evolutionary Approach*, Oxford University Press, Oxford.
Quine, W. V. O.: 1969, 'Natural Kinds', in *Ontological Relativity and Other Essays*, Columbia University Press, New York, pp. 114–38.
Ruse, M.: 1979a, *The Darwinian Revolution: Science Red in Tooth and Claw*, University of Chicago Press, Chicago.
Ruse, M.: 1979b, *Sociobiology: Sense or Nonsense*?, D. Reidel, Dordrecht, Holland.
Ruse, M.: 1982a, *Darwinism Defended: A Guide to the Evolution Controversies*, Addison-Wesley, Reading, MA.
Ruse, M.: 1982b, 'The Revolution in Geology', in D. Asquith and I. Hacking, (eds.), *PSA 1978*, p. 2.
Ruse, M.: 1983, 'Darwin and Philosophy Today', in D. Oldroyd and L. Langham (eds.), *The Wider Domain of Evolutionary Thought*, D. Reidel, Dordrecht, Holland, pp. 133–58.
Ruse, M.: 1984, 'The Morality of the Gene', *The Monist*.
Smith, N. K.: 1941, *The Philosophy of David Hume*, Macmillan, London.
Smith, N. K.: 1923, *A Commentary to Kant's 'Critique of Pure Reason'*, 2nd ed., Macmillan, London.
Sober, E.: 1981, 'The Evolution of Rationality', *Synthese* **46**, 95–120.
Toulmin, S.: 1972, *Human Understanding*, Oxford University Press, Oxford.
Whewell, W.: 1840, *Philosophy of the Inductive Sciences*, Parker, London.
Wilson, E. O.: 1975, *Sociobiology: The New Synthesis*, Harvard University Press, Cambridge, MA.
Wilson, E. O.: 1978, *On Human Nature*, Harvard University Press, Cambridge, MA.
Wolff, R. P.: 1960, 'Hume's Theory of Mental Activity', *Phil. Review* **49**, 289–310.

INDEX OF NAMES

INDEX OF SUBJECTS